THE DAWN PATROL DIARIES

OUTDOOR LIVES SERIES

THE
DAWN
PATROL
DIARIES

*Fly-Fishing Journeys
under the Korean DMZ*

James Card

UNIVERSITY OF NEBRASKA PRESS | LINCOLN

The University of Nebraska Press is part of a land-grant institution
with campuses and programs on the past, present, and future
homelands of the Pawnee, Ponca, Otoe-Missouria, Omaha,
Dakota, Lakota, Kaw, Cheyenne, and Arapaho Peoples, as well as
those of the relocated Ho-Chunk, Sac and Fox, and Iowa Peoples.

Library of Congress Cataloging-in-Publication Data
Names: Card, James (Journalist), author.
Title: The dawn patrol diaries: fly-fishing journeys
under the Korean DMZ / James Card.
Description: Lincoln: University of Nebraska
Press, 2024. | Series: Outdoor lives series
Identifiers: LCCN 2024010593
ISBN 9781496234490 (paperback)
ISBN 9781496241092 (epub)
ISBN 9781496241108 (pdf)
Subjects: LCSH: Fly fishing—Korea (South) | Card, James
(Journalist)—Travel—Korea (South) | Korea (South)—
Description and travel. | Korean Demilitarized Zone
(Korea)—Description and travel. | BISAC: SPORTS &
RECREATION / Fishing | HISTORY / Asia / Korea
Classification: LCC SH667.K6 C37 2024 | DDC
799.12/4095195—dc23/eng/20240509
LC record available at https://lccn.loc.gov/2024010593

Designer and set in Adobe Text Pro by K. Andresen.

To DK

Its only land approach, from the north, was bounded by an almost inaccessible mountain and forest region, and by a devastated 'No Man's Land,' infested by bandits and river pirates. When outside governments made friendly approaches, and offered to show Korea the wonders of modern civilization, they received the haughty reply that Korea was quite satisfied with its own civilization, which had endured for four thousand years.

—"Korea's Fight for Freedom" by F. A. McKenzie

If a scholar only stays in one place, he will be blocked by one small river and become narrow-minded and ignorant. He must travel everywhere to see human affairs, the social customs of the north and the south, mountains and rivers, and to broaden his knowledge.

—Hu Yuan, scholar of the Song Dynasty (993–1059)

The sportsman tries to counteract these inexorable losses by searching out new places, new country. This is how we stay one step ahead of hopelessness and despair.

—"The Hunter's Road" by Jim Fergus

Contents

Illustrations

Romanization and Conversion Note

The most curious thing happened while I was living in South Korea: the proper spelling of words completely changed in a very short amount of time. The government adopted the Revised Romanization of Korean in 2000. Romanization is transcribing the *hangul* alphabetic script into Latin script. Before this was adopted, the McCune-Reischauer system was used for a long time, but it was more cumbersome as it used hyphens, apostrophes, breves, and my heavy-metal favorite, the umlaut.

During this period, an enormous amount of money was spent to swap out street signs, highway signs, subway signs, bus signs, and railway signs. Maps. Textbooks. Business cards. Letterheads. Advertising materials. All of it republished or repainted or reprinted. People's names and existing company names stayed the same, but every spelling of every printed word that did not match up with the Revised Romanization system had to be changed over.

South Korea's second city was called Fusan in the old days; later it was Pusan for many years, and under the revised system it became Busan. Taegu became Daegu and Kimhae became Gimhae. For some lucky cities there was no change: Masan is still Masan. Hadong is still Hadong.

Korean words in this book are spelled using the Revised Romanization of Korean system; however, there are two exceptions: when citing or quoting material from historical content, the outdated spelling will be kept true as the original source spelled it out.

The other exception is for my personal quirks and field notations. One of my favorite trout streams is spelled under the Revised Romanization of Korea as Naedae. In my field notes I always spelled it "Nayday" as it rhymes with mayday. I thought this was appropriate because this creek, like many others in the country, is threatened by forces of destruction and degradation.

There are times when the revised system fails to easily express the pronunciation of a Korean word, and one example is the name of an island that is dear to my heart: Geoje Island. It used to be spelled Koje, but with the revised spelling, the first three letters: GEO, would lead a reader to immediately think of geology or geography and thus pronounce the island's name as Gee-oh-je when it should sound more like Go-jay. In this book, I spell the island's name as Koje, for sentimentality and as a rebellion against a minor flaw in the revised system.

South Korea uses the metric system. Since this book is published in the United States, U.S. customary units will be used when expressing measurements. Money is expressed in U.S. dollars to make it easier for the reader to grasp the value of things. When I lived in South Korea, I rolled with a fistful of Sejongs. (A portrait of the fourth ruler of the Joseon dynasty of Korea, Sejong the Great, is printed on 10,000-won bills.) South Korea is a cash-loving society, and many times I received my pay in fat white envelopes stuffed with these bills, which were similar in purchasing power to that of a U.S. $10 bill. On average, over the years—except for some dips and spikes in the exchange rate—one dollar equaled around 1,200 won.

THE DAWN PATROL DIARIES

1   The Red Quest

It was only a few months after I arrived in Tongyeong when an envelope from overseas was delivered to my school. It had been postmarked in South Dakota. Inside were two letters. One was a sheaf of photocopied papers: a handwritten letter. The other was a handwritten letter that had been written recently, with a yellowed newspaper clipping. The photocopied letter was to my grandfather, my grandmother, and my mother. It was dated December 15, 1951.

> Dear Verna, Eldo, Cheryl,
>
> I can't start this letter off without saying thanks for the letter you sent as I haven't got any mail since I left Japan but in about a week or two, I should get some mail. Until then I'll just keep reading the ones I got in Japan.
>
> Well, I'm no longer in the pipeline status in the service I'm now a member of the headquarters and in service of the 65th Engineers of the 25th Tropic Yellow Lightning Division. Our shoulder patch is red with a yellow border and a jagged line of yellow. As of yet I don't know what I will be doing but they all say I'll be in the maintenance shops and they are tents but that's the same in my line. We are about eight miles back of the lines and all we hear now is the artillery. Otherwise, it's pretty quiet around here.

I got to keep the M1 rifle they gave me at the replacement center and one of the guys looked at it when I was cleaning it and he said it was like new. It sure works smoother than before we left the replacement center. They gave us two clips of ammo and we rode in on open tracks and we thought it was terrible until the driver said when you are that close a plane can stray over and if it's not friendly, you don't want anything in the way of getting out and I agreed.

We sure do eat good here, three meals a day and all you want to eat so boy we won't be hungry. The way it started on the train—getting c-rations—we all thought that is what it would be living on but we sure were mistaken. I've got two bandoliers of ammo I found under my bunk and there is plenty everywhere you go. Everyone is very well equipped as much as clothing and personal needs. We get a PX ration every night. It's for a pack of cigarettes or pipe tobacco, candy bars, gum, soap, razor blades, and tooth powder or paste so you see the guys are well taken care of but of course you only take what you want or need but it's all free.

The country here is all hilly like the cactus hills back home— only the hills get bigger and bigger as you move north. Most all the people here are farmers. They have rice paddies in the hollows and terraces on the slants of the hills where they grow other things. They either walk or ride bicycles. Mostly they walk and then to get stuff to town, if they have an ox, they haul it on a four or two wheeled wagon. Otherwise, they carry it on their heads in great big bundles and they walk along fast and never hold the bundles with their hands. They live in shacks about like the rabbit houses Grandpa used to have.

On the way up here we seen Seoul, the capital city, and boy it is a mess. The buildings are all blown up and you can sure see that Uncle's boys have been through here. The funniest thing I saw was one big stone building with only the front of it left standing. The doors to it were closed and no windows were in it. I sure thought it looked odd.

The guys I'm with seem to be very nice and the old guy who I got acquainted with at the replacement center is with me in the HQ but he's in the next bunker. Our bunkers are in the hillside and made of logs and then covered with dirt so it's nice and warm as there are no drafts. We have a stove, electric lights, and a radio set up so you can hear through ear phones— I'm listening to it now. The sergeant in our bunker says I won't be given a job until after we get to where we are going as this outfit is going into reserve for a while. I don't know for how long but we are moving out in a day or so. It's sort of cool now in the early part of the day—otherwise it's not too bad.

I sent Cheryl a hankie in mom's one letter but it wasn't much but we aren't in a position right now where we can send anything but when we can I'm going to send some more stuff. Hope this finds you all well and happy. I'm feeling fine and I'm getting along nicely so don't worry about me as I've got the good keeper on my side to take care of me and I'm sure going to help him do his job. So bye now.

Lots of love and kisses,
Uncle Clayton

I unfolded the second letter and set aside the newspaper clipping.

Dear Jim,

I've been going to write to you for a long time and now finally Grandma found the paper with your address on it so here goes.

Your Grandma called and asked Marty if I ever wrote to you about the places I was at and that really got her looking.

To start off: "My friends and neighbors have selected you to serve in the U.S. Army."

I was sworn in on the 26th of April 1951. I went to Fort Lewis, Washington, then across the country to the Aberdeen Proving Grounds and six weeks of basic training and then to Atlanta,

Georgia to the ordnance training school and of course they made a welder out of this farm boy.

The school lasted ten weeks, then fifteen days of furlough time at home and I was on my way to Seattle, Washington and Pier 91. We sailed out to Japan and to Korea on the Marine Adder. We were in Japan for thirty-six hours and were assigned to our units in Korea. Back on our ships and more big waves and bobbing up and down.

We had to anchor way out in Inchon Harbor as the tide was out, so we got one more cooked meal on the boat and got in closer to shore and climbed down the side of the ship on a net into the landing barges. They ran aground and dropped the end down and everyone sloshed out through the mud and water. We marched in rank and loaded into narrow-gauge box cars that had wood bunks, or more like shelves to lay on when all the guys were loaded, we went to the replacement depot.

There we slept in an old hospital building if my memory is right, it was like four stories tall, just the four walls and window holes and the bottom cement floor was like sleeping on ice. They said no lights of any kind flashlights or lighters or smoking. Then we got a can of c-rations to eat—hot on the outside, still froze in the middle (I've got my plastic spoon I ate that with).

Grandma took a map somebody sent me out of the Sioux Falls paper and is getting a copy made of it. I drew lines with arrows of where I went and the places we were stationed at and worked from are circled. There are lots of areas that had names that aren't on the map like The Fingers (hills like a hand), the Iron Triangle, Jane Russell, Sandbag Castle and the Punchbowl. We did a lot of work and war there. We built supply roads and bridges in two directions out of the Punchbowl, both to the front lines and out from there.

My sidekick on these bridges was Jensen, a Swede from Iron Mountain, Michigan. He'd always weld "Red Jensen" on one end of the bridge and the date of completing it. One of these, C Company 65th combat engineers built when it was minus ten below zero when we welded it together in open water in the middle of the river, the name of it I don't remember.

Along the ridgelines of the Punchbowl was a road called Skyline Drive. It was a bad road to travel as some of it you had to drive on top in plain sight and it wasn't good enough to drive fast on so you always got some "stones" thrown at you in those spots = boom! and dirt and rocks raining down on you and your welding truck. It was always said if you hear the noise and feel the rocks you are lucky.

The coldest temperature I seen there was minus thirty-five below zero. It froze our diesel fuel we used in our tractors and tent stoves solid. I got ours burning by mixing gas in it—very dangerous to do. In times like that a tent with one small stove is very cold to live in. The hottest temperature I seen there was 112 degrees. Three of us took turns cutting with a torch making some big pulleys for some supply line tramways for reaching the hilltop trenches and for bringing down wounded guys. You'd cut till your colored glasses filled with sweat and then hand it to another guy.

If you've got a map, you can compare it to this one and you may find some of the towns.

Love,
Uncle Clayton

It was from my great-uncle on my mother's side of the family. I remember playing at his farm in South Dakota when I was a boy. I rode horses, blasted fireworks, and shot many guns. The first semiautomatic weapon I ever fired was his Remington Nylon 66. I spent an afternoon picking off gophers with the rifle in his pasture. I remember him with great fondness.

I looked at the map from the old news clipping and matched it up with a map that I kept on my desk. The Korean War was nicknamed the "Forgotten

War." It was overshadowed and bookended in American consciousness by World War II and the Vietnam War. No peace treaty was ever signed. There was only an armistice, as if a pause button was pushed. The rifles were never put down. The area where my uncle served is the northernmost region of South Korea, and I lived in a small city on the far southern coast. Somewhere up by the DMZ were bridges that my uncle had built as a combat engineer during the Korean War, and a guy that he worked with had welded his name on these bridges. I vowed to find one of them.

2   Archipelago of Sword and Spear

The school that sponsored my work visa provided me with a small apartment on an island that was connected to the mainland by a bridge and an undersea tunnel with the inscription: "Dragon Gate Leading to Sanyang." The apartment building was part of a larger complex, and my building was on the far end. I was on the fourth floor, and the balcony on the back of the building faced a mountain covered with a lush forest. I studied it every day, looking over the contours and ridgelines and wondering what was up there.

After my first week of classes, I set out to climb Mireuk Mountain on an early Saturday morning. I hiked toward the patchwork of gardens and farm plots and picked up a small cement road that led up into the foothills. Eventually the road ended, and there was nothing but a dirt path. The garden patches were smaller, and there were farming implements lying around: shovels, rakes, buckets, and irrigation hoses. I smelled a garbage fire in the distance.

I stopped and looked around. I wondered if I was trespassing on private property. There were no Keep Out signs or Private Property signs. Not that I would have understood them anyway. I had gotten off the plane a week before and had memorized only a handful of expressions from a Lonely Planet phrasebook.

Nobody was around. I hung there and pondered my next move. I looked back at my apartment building. I counted the floors upward until I found my balcony, and I memorized the spot as a reference point. On the elevator control panel there was no button for the fourth floor. There was one button with the letter F. The word for four and death, *sa*, are identical. It is an unlucky number, to be avoided at all costs. These were the cheapest apartments.

Where I grew up, in rural Wisconsin, I roamed throughout the coulees, free of any hassles, with a lever action .22 rifle and a bow and arrow. Later, when I was a teenager, more of the farmland was purchased and subdivided into smaller rural lots. People got a lot more sensitive about where an armed teenager could roam around. No Trespassing signs were posted. I respected the property lines but the freedom to roam of my boyhood was gone.

I looked up the dirt path and decided to keep hiking. If I got hassled, I would handle it. I wasn't out to cause trouble, and I was only passing through. The trail entered the woods and it was well worn. The clay was compacted, and, over the years, the rocks and dirt had merged into steps compacted by footfalls. I passed an old man, who greeted me with a smile and said something I didn't understand. He wore a backpack made of forked tree branches with woven twine for shoulder straps.

That moment of wondering if I was trespassing on private property was the start of an important insight that would serve me for the years that I would live in South Korea. The insight was to be cultivated through study, discussion, and time spent afield. Once this insight was better developed through the study of history, farming, and the politics of land use, it was a ticket to explore all of South Korea with total and complete impunity. I did not know it at the time, but this would be the key element of my fly-fishing endeavors. I did not need to ask anyone's permission: just go. Go wherever and whenever the spirit of exploration called.

There were signs farther up on the mountain. I copied the Hangul words down in my field journal and presented them on Monday morning to Mrs. Kim, my co-teacher at the school. She laughed. They were not No Trespassing signs but of the more informative kind: this trail leads here, there is a spring up ahead, and the Buddhist temple is that way.

I studied. At the bookstore, I purchased Hangul handwriting workbooks to learn the Korean alphabet. They were the same ones used by kindergarteners. I traced each of the fourteen consonant letters and ten vowel letters and learned how to put them together to make words. Words like *apple* and *dog* and *chair*. As I wrote them, I pronounced them.

I made stacks of flash cards to develop my vocabulary. I kept a deck in my pocket and would pull them out in times of boredom. My personal test was to be able to read street signs without pause. The best way to do this was on the bus. The faster the bus, the harder the test. The challenge was to be able to read a sign at a second's glance. Sometimes I did not know the meaning of a word, but I could read it and pronounce it properly.

At the school office, I traced a map of South Korea—just a silhouette of the peninsula—and made photocopies. These were worksheets. The test was to populate the map with Korea's major cities. I wrote the name of each city in Hangul, and I also drew borderlines denoting the provinces. Some of the cities in the Gangwon Province matched the names on the DMZ map that my uncle had sent me. I did not care about the places surrounding Seoul in the Gyeonggi Province. Seoul is a primate metropolis, and the outlying dull and subservient satellite cities all blended into one giant mass of urban sprawl. Nor did I have much interest in Jeju Island, Korea's honeymoon destination, which was a semi-tropical landscape of great beauty—and therefore mobbed with tourists.

After memorizing the locations of cities, I also sketched out the locations of Korea's major rivers and mountains on my worksheet maps. All of these places seemed very far away. I was in Tongyeong, a land's end fishing port on the southern coast.

I did much of my studying at *dabangs* (old-school coffee and tea rooms) at a time when more Western-style cafés were popping up throughout the country. Dabangs were on their way out; Starbucks-influenced shops were in. Dabangs evolved from the fin de siècle days during the late Joseon Dynasty, when Korea was opening to foreign envoys, and later became a place for artists, writers, and poets to gather. As Korea modernized, they were more often used as meeting places with business associates—a place to meet someone and have a smoke and a coffee. Many of the ones that were left were "ticket dabangs" and were fronts for prostitution. The giveaway is mini-skirt-wearing girls leaving the establishment carrying serving trays, each tray wrapped with a small tablecloth that holds a thermos and cups. Parked outside the dabang is a young tough with a motor scooter waiting to deliver her to a called-in address. She hops on the back of the bike and rides side-saddle with the serving tray on her lap.

This was where I studied Korean. There was one near my school that served the strongest coffee around. The madam never changed the coffee grounds. She was friendly, and the girls that worked there thought of me as a curiosity. They would join me in a booth and ask questions in broken English. I was studying there one afternoon when a loud air raid siren went off across the city. I looked out the window, and people were scurrying indoors and vehicles were pulling off to the side of the road. The bustling street scene was emptied in seconds. I asked them if North Korea was attacking. They laughed. It was a civil defense drill.

These pretty little hookers squealed with amusement when I mispronounced a word in a goofy way and then quickly affected a schoolmarm air when correcting the mistake. These young ladies were what Korean men referred to as "nine-tailed foxes," flirty temptresses that, according to a folktale, lead men to their destruction. One of them, Mi-na, wore an armband to hide cigarette burns.

I also studied there because it was always warm. There was no central heating in the school and no insulation. Electric heaters were wheeled about and hauled around by their power cords like dogs on a leash. It was a place where people ate lunch and studied with their winter coats on. My school was not an exception; all public buildings were like this. Another relief from the chill was drinking *yooja* tea made from citron oranges. It tasted like hot lemonade. At one time a grove of *yooja* trees could pay for tuition for a four-year college. It was nicknamed the university tree. Farm co-ops would send out workers to pick the fruit for you and hand you a check. It was overplanted and the market crashed.

I lived on Miruek Island and commuted over the bridge by bus to teach in Tongyeong. The city has a heritage of fishing and warfare. The port is considered the most scenic in the country and has one of Korea's largest commercial fishing fleets, including many squid boats strung with hundreds of light bulbs for night fishing. Fishing boats were triple moored side by side. The ships had eyes painted on the hulls, and leafy bamboo shafts fifteen feet high were lashed to the flagpoles. The boats ventured into the Hallyeo Haesang National Marine Park, a coastal ecosystem composed of hundreds of mountainous islands (both inhabited and uninhabited), sand beaches, gravel beaches, jagged capes, and rugged promontories. Mireuk Mountain—right in my backyard—overlooked all of this. The mountain had been an observation post for centuries. The 1,512-foot peak became my playground.

Every weekend I set off with my pockets loaded with baked sweet potatoes and *Chungmu kimbap* rice rolls. I carried an empty canteen, which I would fill later at a natural spring. The forest on the mountain was composed of Japanese cedar, paulownia, black pine, bamboo, and camellia. I made my way through the trails and studied the Buddhist artwork and traditional carpentry at Mirae Temple and Yonghwa Temple. Down by the waterfront near Life Town apartments, I got to know a boat builder who had the skeletal wood frames of fishing boats up on blocks. He used traditional hand tools, and every week I stopped by to check on his progress.

When I had to run errands in the city, I stopped at Gangguan Harbor and the Jungan Fish Market. The smell of seawater was clean and rich until I reached the market, where the air was heavy with garlic and fresh fish. I wrote down the Korean names of the fish and sea creatures I had never seen before and later translated the names into English. There were many types of sea bream, squid, octopus, sea squirts, anchovies, abalone, oysters, mussels, clams, seaweed, sea cucumbers, and penis fish. There was yellow corvina, a favored fish used in Confucian ceremonies that honor the dead. I found a fish monger who always had a pot of *maeuntang* simmering. She served up the spicy fish stew with a bowl of rice and kimchi and a side of grilled mackerel.

Nearby Hansan Island was the base camp of Admiral Yi Sun-sin, the naval warrior who defeated the Japanese navy numerous times during the Japanese invasions of Korea of 1592–1598. It was right in these waters that he ambushed and sank 141 Japanese ships with a fleet of twelve armored "turtle" ships. I visited the memorial there and spent other weekends island hopping to the islets of Bijin, Saryang, Yeonhwa, Yokji, Somaemul, and Maemul.

And then the year was over. I had never made it up north near the DMZ to look for the welded signature of Red Jensen on some lost bridge built by him and my uncle during the Korean War. I hadn't even made it to Seoul. I had been to Busan twice to exchange money at the black market and to buy books and get a cheeseburger at the Seaman's Club on Pier 8 and to party on Texas Street. With a girlfriend, I had hiked some trails in Jirisan

National Park for a weekend. I had spent very little time on the mainland. I felt like I had barely gotten to know the country.

I was deer hunting back in Wisconsin when I got a phone call. It was from my old college friend John Erickson. He was still in South Korea, partway through a one-year teaching contract. He passed on a tip: there was a job opening on Koje Island, about thirty minutes down the road from Tongyeong. It was a larger island I had never gotten around to visiting as I had been too busy exploring the nearby tiny islands. I remembered hearing good things about it.

THE SWORD

Every weekday morning at six o'clock I bowed before Master Hwang. The *dojang* was located by the waterfront, not far from the Samsung shipyard in Gohyeon. It was unheated in the winter, and the wooden floor was always cold. We meditated and then rose, swords at our sides, and began the striking and slashing repetitions to form muscle memory. We counted each rep in Korean. I was a student of *kumdo* (the way of the sword). It was the Korean cousin of Japanese kendo. I wore a heavy-woven indigo-blue jacket and pants that were loose and billowy to allow for maximum movement.

Before sparring, I knelt in front of my armor, which was stacked near the wall. I tied a cotton bandanna around my head that would keep sweat from my eyes. A kumdo duel is an aristocratic waltz of ultraviolence. Each opponent slide-steps forward and backward at gliding angles while smashing each other with bamboo swords. Once the helmet is on, there is no way to wipe your eyes or nose or face. I once sparred with Master Hwang while I had a cold. He toyed with me but I defended myself well. I could not breathe from all of the mucous clogging my nasal passages, but it eventually came out as I gasped for air—and I spent the last minute of the fight with my face covered with foamy, bubbling snot.

I put on the *gapsang*, a thick cotton belt made of three hard canvas flaps that protect the waist and groin area. Over that was the *gap*, the chest protector made of lacquered bamboo with a shark-skin finish. I put on my helmet and fitted it tightly against my chin. I saw the world through a metal grille. Nothing could hurt my head or my face in a frontal attack. I wrapped the cloth string around the back of my skull and knotted it off. I put on my mitt-like gloves, which were well padded in the hands and wrists, but the forearm area was only thick canvas and getting hit there always stung. I grabbed my *juk-do* (bamboo sword). It was made of four bamboo slats bound with leather and a tensioning string. I stepped over to the training dummy, a wooden post with chunks of sawed-up car tires.

There are four main targets in kumdo. The main strike is to the top of the head, never horizontal like a Hollywood beheading. Another is a slashing movement that cuts from the upper torso to the lower torso at a downward angle. The third is a downward strike to the wrist of the strong arm guiding the sword, and the fourth is a piercing jab to the throat, propelled by a stutter-step lunge. If such movements were used with a real sword, the opponent's head would be split like a melon, he would be eviscerated with his entrails spilling to the ground, his right hand would be chopped off, and there would be a hole where his Adam's apple used to be.

After practice, I drove back to my apartment on my scooter, a 49cc Hyosung that ran like a chainsaw with two wheels. I got cleaned up and went into school for an early morning class. It was a class of doctors who were old friends. There were only four of them, and they used the class mostly as an excuse to get together. Keeping their language skills sharp was a bonus. All of them spoke English well enough to have interesting conversations.

They told me about their weekend plans: they were chartering a fishing boat and scuba diving near a rocky islet off the island's southern coast. When I told them that I had my PADI certification, they invited me along and drew a map to a dive shop in town where I could rent the gear.

The fishing boat turned out to be a simple wooden vessel, but the captain had added an open deck area for divers to put their equipment on, and he had added a ladder for getting back into the boat. I had gotten scuba certified through a college course, and besides the swimming pool, the only diving I had ever done was at Pearl Lake in South Beloit. After passing the skills exam, I swam into a sunken school bus and watched a school of yellow perch glide through the windows. This would be my first time diving in salt water.

We sailed to Hongdo, twenty-seven miles south of Tongyeong. It is called Egg or Seagull Island as so many black-tailed gulls nest in the jagged cliffs that the island looks as if it is capped with snow during the warm months.

The water was clear blue, loaded with fish and heavily oxygenated by the waves smashing against the rocky islet. I noticed one of the doctors was swimming with a stick. I watched him. It was a spear. When we took a break aboard the ship, I asked to borrow the spear. He showed me the rubber band contraption that shot the spear forward, and he gave me a net bag. I went back into the water. The crevices and underwater nooks of the islet held many blue crabs, and I hunted and killed them one after another. It was satisfying to see and hear the steel of the spearhead smack into the shell of the crab. I hunted until the bag was full. I dumped it into the boat, shed my nearly empty tank and breather, and flopped back into the water with just the mask and snorkel to kill another bagful.

At the end of the dive, we had lunch on deck, and we feasted on crabs and sashimi. They nicknamed me "crab killer." I stripped off the wetsuit, and the sunshine warmed my cold, salty skin. I was feasting and drinking with new acquaintances on a boat in the South China Sea next to an uninhabited island. I had found my calling.

During this period, I was reading and rereading the great Asian works of warfare and combat: Sun Tzu's *The Art of War*, Miyamoto Musashi's *The Book of Five Rings*, Yamamoto Tsunetomo's *Hagakure: The Book of the Samurai*, and Takuan Soho's *The Unfettered Mind: Writings from a Zen Master to a Master Swordsman*. Also, my reading delved into traditional Korean poetry written by rebels exiled into the hinterlands, and I also studied *Son* (Zen) Buddhism. Peter Matthiessen's *Nine-headed Dragon River: Zen Journals 1969–1982* was a good introduction to this religion. I also amassed a collection of books about the Korean War.

On the weekends, I used my scooter to explore the island's interior backcountry, including the ancient fortresses of Saedeung, Oryang, Oksangeum, Gohyeon, and Pyewang, on the island's west side. All that remained were rock walls overlooking key defensive high ground on the island. At the time I had a poem by Kim Chongso copied into my field journal:

> North wind bitter through the branches,
> Bright moon frigid in the snow,
> Great sword drawn, I stand in this remote border fortress.
> Long whistle,
> Mighty shout, nothing dares oppose me.

On the northern end of the island, I found a small memorial for Dr. John R. Sibley, an American surgeon who provided medical care on Koje in the late 1960s and 1970s. Some evenings I would grab a few beers and hike over to the remains of the Koje prisoner of war camp. All that was left were some crumbling cement walls, and I would drink and read until the sun went down. A memorial park for the camp was being built up the road, but these old walls were the last vestiges of the original compound.

"Until the war was well advanced, Koje-do remained one of the prettiest possessions of South Korea. A little fishing community a few miles across the sea from the port of Pusan, it had gone untroubled by most of the dramas and horrors that had befallen the mainland," wrote Max Hastings in his book, *The Korean War*.

In literature about the Korean War, there is always mention of refugees heading south. Refugees fled as far south as they could until they hit the ocean, and then they migrated farther south into the coastal archipelago. On Christmas Day in 1950, the U.S. merchant vessel ss *Meredith Victory* delivered 14,000 Korean refugees to the safety of Koje Island. They had boarded the ship at the northeastern North Korean port of Hungnam as the Chinese army bombarded the port and advanced on the city. It was a three-day passage through mine-infested waters, and the ship carried 3,000 barrels of jet fuel below deck. Five babies were born during the trip. Nobody died. Captained by Leonard LaRue, it is considered to be the greatest rescue operation ever by a single boat.

Other than refugees, Koje Island mostly avoided the war until it was selected to become a prisoner of war encampment. At first, Cheju-do, Korea's honeymoon and tourist island, located way off the mainland, was considered as a place to house the thousands of prisoners taken by UN forces. That idea was nixed and Koje was picked.

"Thereafter a decision was made to transfer the prisoners to Koje-do, a much smaller island just a few miles southwest of Pusan. This was merely a choice between evils, for Koje-do was itself hardly the ground a sane man would have chosen to erect camp sites. It was rocky and mountainous, with almost no flat ground for construction and proper dispersal of the compounds. As a result, Koje-do was very soon crammed full with far more humans than nature had ever planned for it to support," wrote General Matthew B. Ridgeway in *The Korean War*.

Just as Ridgeway surmised, turning Koje Island into a prison camp did not work out well. The camp quickly became overpopulated. In 1952 there were over 170,000 prisoners. Mass demonstrations and riots erupted. There was bad-blood viciousness between the South Korean guards and the captured North Korean soldiers. Some of the prisoners had allowed themselves to be captured as they were agents with a mission to infiltrate the camp, organize riots, and enforce Communist doctrine. Kangaroo courts were established inside the camp, and beatings and murders with homemade weapons were common.

During one uprising, outnumbered guards opened fire on an armed mob and killed fifty-five prisoners. On May 7, 1952, the Communists captured camp commandant Brigadier-General Francis Dodd, stashed him in a tent, and presented a sign: "We capture Dodd. As long as our demand will be solved, his safety is secured. If there happen brutal act such as shooting, his life is danger." They put Dodd on trial for alleged brutalities while American forces brought in reinforcements, flamethrowers, and tanks. Three days later, they released Dodd after many negotiations. It was a political propaganda victory for North Korean and Chinese Communists.

I contemplated that I was about the same age as the soldiers who served in the Korean War. A man could be drafted and killed, maimed, or thrown in prison. Life is short and sometimes ugly. All of this reading and study of history made me think about my life and where I wanted to be. It was time to act.

I asked out a Korean woman I was interested in. One of my coworkers was from Newfoundland, and he was helping a Korean elementary school teacher with a project. She was very attractive, and my coworker was infatuated with her. The project worked out well, and I learned they were meeting at a fancy coffee shop to celebrate. I stopped by out of curiosity and pretended it was a coincidence. He introduced me to the elementary school teacher. She was beautiful and elegant and way out of his league. It was obvious she was having coffee with him as a nicety for helping her out. I exchanged hellos and left. A day later, I rifled through his address book and found her number. I called her and asked her out. We started dating, and when the Newfie learned of this double-cross, he never spoke to me again. The second decisive action was breaking my teaching contract. I lined up a new job at the Daewoo shipyard that involved teaching English to engineers. I would nearly double my salary, be provided a company car, and be required to wear a gray long-sleeved work shirt and gray work pants that bloused over a pair of steel-toed work boots. From executives to welders, everyone wore the same outfit. Breaking my contract created much drama at the school, and there were meetings, negotiations, and passive-aggressive bitching. I held fast and soon I was moving to the small city of Okpo, a short drive down the road and home to Daewoo Heavy Industries, one of the biggest shipyards in the world.

THE SPEAR

I swam by a rock that was looking at me. I paused. The rock had eyes. The rock had arms. It was a long-arm octopus (*Octopus minor*). It was camouflaged well. I cocked back the spear and shot it in the bulbous head. The eight arms came to life and wriggled with a death spasm and grabbed the shaft of the spear. It changed colors, and there were iridescent hues of chartreuse, gray, navy blue, and brown. I surfaced and rested the impaled creature on a rock, and when it stopped writhing, I stretched it out. I found its beak and looked at the suckers. The octopus was aptly named. The head was quite small compared to its arms, which measured around twenty-three inches against my pole spear. I had never cleaned an octopus before. Down at the seafood markets, I had watched the fish wives scour their slimy carcasses on wood cutting boards using rock salt as an abrasive. I shoved the tip of my dive knife into its beak and made a slit to the top of its head. Then I removed the entrails, brain, and other parts and flung them onto the rocks to feed the tiny beach crabs. I rinsed out the insides with seawater, folded it up, and put it into my net bag with the tentacles still twitching.

The closest comparable blood sport to spearfishing is bowhunting with a recurve (or stick bow) on the ground. The key parts of the comparison are being on the ground and the use of a recurve. With a pole spear, the shaft of the spear must be held in the hunter's grip until the moment the shot is taken—much like a recurve bow. This is opposed to a compound bow, where there is a mechanical advantage as the draw weight is reduced greatly through the action of cams. A hunter can hold back that power for a long time. The other comparison is being on the ground or, in spearfishing, being on the same level as the quarry. The hunter and the hunted are eye to eye, unlike a bowhunter using the height advantage of a tree stand. Being on the same level as the prey requires stealth and patience and calculation.

The pole spear had a two-piece fiberglass shaft that screwed together in the middle for a total length of six feet. A large loop of rubber tubing attached to the end, and a three-pronged spearhead screwed in on the business end. I swam with the band looped, slightly tight, around my hand, like an archer's fingers tight on a bowstring, ready to draw. Swimming with the rubber loop stretched out exhausted my grip strength. Once I spotted a nearby fish, I stretched the band along the shaft with both hands until it was tight—essentially "cocking" the weapon. I extended the spear toward the target fish with my grip hand and let go. The spear would shoot out about one spear length depending on how much stored energy I gripped with the stretched rubber band.

At close ranges the spear prongs penetrated the fish; farther out was risky and only the tip might pierce it. The tips were barbed but not much, and the fish could easily wiggle off and swim away injured. In that case I always tried to pin the lightly speared fish against a rock and then ease down along the spear shaft and pluck off the fish with one hand while holding it tightly. This dulled the spearhead, as did errant shots that smacked into rocks—and the fish were always in the rocks. I kept a mill file in my dive bag and touched up the points on the beach after every dive. The fish went into a net bag with a spring-loaded mouth opening and a drawstring cinch at the bottom for dumping out the day's catch.

After the first dive trip with the doctors, I returned the equipment to the dive shop and immediately purchased a mask, snorkel, fins, wet suit, pole spear, and net bag. I had a pair of camouflage neoprene gloves from my duck hunting days, and they worked perfectly as dive gloves. I swam in the East Korea Warm Current, a tributary of the Tsushima Current. Like the Gulf Stream, it is one of the great ocean rivers of the world. It flows along the southeastern coast of the Korean Peninsula and meets the North Korea Cold Current.

The water was abundant with sea life, and that made me feel rich—that a great dinner was out there for the taking. You just had to swim for it. I felt rich overall. I was mentally and physically fit from my daily swordsmanship training. The new job had long hours, but driving from class to class within the shipyard I would roll by ships the size of horizontal skyscrapers. I passed the massive goliath crane, the gantry cranes, and the jib cranes. I passed the quays and dry docks holding Aframax tankers, RORO vessels, LNG carriers, and floating production drillships. I would listen to my students talking about their work and, in return, I got a free education about the global shipping industry. There were also heavy machine guns mounted on top of the buildings. The shipyard was a strategic military target, and there was one section for building military vessels.

My Korean girlfriend was an elementary school teacher in the village where she grew up. She was an island girl. We were both busy during the week, but the weekends were ours to share. I taught her to drive a stick shift in my work car, and we practiced in the parking lot of an abandoned amusement park that overlooked Okpo. Sometimes she would come to the beach with me and lay out in her bikini while I spearfished for our dinner.

But mostly I spearfished alone. I liked to get up before dawn and make some coffee and drink it out on the beach as the sun came up. Those cool mornings were what I lived for after a work week filled with long hours. I liked stripping down to my shorts and letting the cold wind brace my skin as I pulled on the wet suit. I liked spitting into my mask and rubbing the saliva around as a natural lens defogger. But most of all I loved killing fish with a spear.

There was surf perch, small bluegill-sized fish that were common but challenging targets. My favorite were striped beakfish (*Oplegnathus fasciatus*), also called barred knifejaw. I called them zebra fish. They were crappie-sized panfish with black stripes upon silver bodies. These were always among the rocks, and some days the weather would get rowdy and waves would crash me into the rocks as I was getting into position to take a shot. There were black sea perch (*Sebastes schlegelii*) and black sea bream (*Acanthopagrus schlegelii*), two fish that were highly regarded among sashimi connoisseurs.

There were four-striped grunters, which I nicknamed "stripeys," and spotty-bellied greenlings (*Hexagrammosa grammus*) that lurked among the rocks. For the sandy beaches, the only game in town was hunting for flatfish. At first, I wrote off the sandy beaches for two reasons: they attracted crowds in the summer months, and they lacked structures (rocks) that fish are attracted to. Then I killed my first flatfish. The first thing you notice are the two eyes atop the dinner plate–sized body that matches perfectly with the sand. It was relaxing to swim over large expanses of sandy beaches looking for the hidden prey and, once spotted, it was death from above.

Between the rocks and the sand were mullet. I nicknamed them gray ghosts. To see one straight ahead was rare. It seemed like I could spot them only out of my peripheral vision, and then they were gone. If I did get the jump on one, it was always a moving target. I never once observed one hanging in the water column. After many misses and wild shots, I remembered a lesson from goose hunting: lead them by at least a body length. I led them by three body lengths, and eventually my spear tip punched mullet flesh.

Pufferfish (also known as swellfish, *fugu* in Japanese, and *bogeo* in Korean) were abundant and easy to kill. I shot them only if they came close to me because I didn't want to be near the poisonous little bastards, even though they were harmless. They are dangerous only if you eat them, and every year across Asia a few people die from the dangerous delicacy. The fish contains a neurotoxin in its internal organs that guarantees death, yet some gourmands ask the chef to leave a trace of the poison as it provides a numbing sensation to the palate and a pleasant buzz. There is a joke that patrons freeze at the clink of steel chopsticks hitting the floor at a *bogeo* restaurant. I shuddered at the idea that a speck of poison on my spear prongs could contaminate the other fish that I would later fillet. After shooting a puffer, I stabbed my spearhead into the sand multiple times to scour the surface.

I slaughtered jellyfish whenever possible. It was payback for the times I was stung while swimming the beaches without a wet suit. They move into coastal waters when the water temperature goes up in the summer. At first, I thought urine would ease the pain and the itchy burning. Pissing down the inside of your leg is simple; trying to piss on the back of your leg is difficult. You end up pissing all over yourself. I later learned that urine as a jellyfish-sting treatment is a myth, and the best treatment is heat. The remedy was always the same: quickly build a driftwood fire and heat the blade of my dive knife. Pressing the warmed blade flat against the sting eased the pain, and with the edge I could flick off any tentacle tissue. Reheat as necessary.

The culprit was Nomura's jellyfish (*nemopilema nomurai*), a monstrosity that can grow bigger than a person and weigh over 400 pounds. They live in the waters around Japan, Korea, and China and grow big from devouring zooplankton. I never saw a large one. The ones I killed were about the size of a basketball or smaller. I aimed for the translucent mushroom-cap body and smacked it dead center. Once it had been impaled, I swam over to a rocky outcropping and scraped it off my spearhead. I left them to desiccate in the sun and feed the seagulls.

On the outskirts of Okpo, I always stopped at a hole-in-the-wall restaurant to get a bowl of *naengmyeon*. It was a dish that I despised when I first tried it. It was not so much the taste as the temperature. At first glance it looks like noodle soup, but it is served cold. I remember eating it and wishing for a microwave. I forced myself to try it again many times, and later I learned to love it. It was refreshing and satisfying on hot summer days after spear-fishing. The recipe originated in North Korea. It is very simple: buckwheat noodles served in cold beef broth and topped with a hard-boiled egg and slices of cucumber and beef. Add in a dab of red pepper paste for some zing. The trick to eating it is to snip the noodles into smaller lengths with kitchen shears, as the buckwheat noodles are long and stretchy.

The weather on the island was mild enough that I could swim into late November before it got too cold. Okpo put me on the eastern side of the island, and I had a portfolio of beaches and rocky points that I would hit on weekends. Guyeong Beach and Nongso Pebble Beach were on a northern peninsula, and Heungnam and Deokpo were farther down the eastern shore. As I explored further, I found I preferred the southeastern coastline. Gujora Beach was a favorite, and while swimming there one night I was surrounded by bioluminescent algae. I did the backstroke with the starry sky above me and my body surrounded by sparkling water. It was such a surreal experience I ended up over a mile offshore and had to sidestroke my way back to the beach.

Hakdog Pebble Beach was a good spot, but it was the rocky island of Haegeumgang, on the southern tip of Koje Island, that got my attention. Designated as a national scenic site, the island is composed of massive sea cliffs and is named after Geumgang (Diamond) Mountain in North Korea, therefore meaning Diamond of the Sea. Sightseeing boats cruised Haegeumgang, so to avoid them I focused on rocky promontories in the area. Using some climbing rope, a Swiss seat, one carabiner, and a figure-8 descender, I rappelled down to a small rock ledge near the water. This spot was the only time and place where I spearfished and snorkeled with a sense of fear and dread. Sharks are present in Korean waters yet rare enough that it makes the news when a fisherman snares one in a net. Most reports of them come from the Yellow Sea off the western coast and the southern Cheolla province. A male diver was killed over there by a great white in 1996. Here, the water was very clear with a rich light blue color that faded into the deep. It was the deepest water I'd come across, and it felt like I was swimming over a bottomless void from which anything could emerge before I could react. I spearfished there once and never returned.

One morning I drove to the shipyard, and the main gate was blocked by hundreds of protesters. They were union guys: the welders, steel cutters, forklift drivers, and crane operators who were the backbone of the island's economy. They wore headbands and waved banners and flags with the letters "IMF." They pounded on *buks* (barrel-shaped drums) and *jang-gus* (hourglass-shaped drums) and were chanting and making a hell of a racket. I turned around and went to another gate. It was blocked, too, and so was another one. There was one gate near the education and training building with no protesters. The gate was locked, so I parked the LeMans and climbed over the fence.

The atmosphere in the office was tense and abnormal. Some of the office workers were looking out the window at the empty shipyard and watching the protesters off in the distance. This was the crack in the Asian Economic Miracle. I had been reading of the meltdown in the newspapers, but now it was real. The rapid industrialization and exploding GDPs could not be sustained by the Asian tiger economies and their dirigisme directives. South Korea's *chaebols* (family-owned conglomerates) were in such haste to build global enterprises that the country's banks were burdened with deadbeat loans for unprofitable and dubious ventures. In the news there were corruption charges aimed at government officials and corporate executives every day. There had been way too much easy money floating around over the past couple of decades, and it all had to catch up sooner or later.

The South Korean won collapsed, going from around 800 to 1,700 won per U.S. dollar. The International Monetary Fund (IMF) was stepping in with billions in bailout money, but the catch was that corporate Korea had to restructure its business affairs. Everyone knows that when used in the workplace, the word *restructure* means people will lose their jobs. This is why the blue-collar union guys were protesting. Their work at the shipyard was all they had. Like me, they had built a good life on the island and did not want it jeopardized. The island was flush with cash with the

Daewoo and Samsung shipyards employing thousands of workers in solid blue-collar and white-collar jobs. Orders for new ships came in from all over the world, and that brought engineers from all over the world. Scantily clad women danced on the sidewalks and sang karaoke songs in front of newly opened stores that serviced the booming island economy. This was the most drama the island had seen since 1998 when a North Korean semi-submersible spy boat was sunk after a seven-hour chase sixty-two miles off the island's southern coast.

I met my supervisor, Mr. Lee, and he approached me like a nervous process server. We talked about the economy and the headlines, and he mentioned that if there were job cuts, some of them might be in this department. Some of the training programs might get axed, including the English program. His wife had just had their first child, a baby daughter. He was worried about his future.

South Korea's credit rating plunged, and the Seoul stock exchange was a bloodbath. The collapsing economy drove the interest rates up, and some Korean companies begged for emergency loans, while others ditched assets at fire sale prices. Overseas investment was opened up, sparking xenophobic paranoia of Korean-owned institutions being overtaken by foreign vultures.

I took stock. My paycheck had lost half of its value. I had to cover my student loan debt, but I was ahead on my payments and had money in a stateside account that could service it for a while. For many years, South Korea had been the kind of place where an English teacher could roll in, work hard, and make enough cash to pay off student loans, fund some world travel, and save up enough seed money for life's next big step—graduate school, a down payment on a house, start-up money, whatever. Now, with the collapsed exchange rate, the money made here was no better than a low-value Third World currency, but as long as I didn't need to trade my Korean won for American dollars, everything was pretty much normal.

Mr. Lee told me that classes were canceled for the week. He said I could work at my desk and prepare lesson plans. I had never worked at my desk before—I was always in the meeting rooms throughout the shipyard teaching classes—but I understood that Koreans liked to have warm bodies accounted for in the office. I asked if I could work from home. I said that the protesters made me uncomfortable, being a foreigner and all. This was not true. I got along very well with those guys. He made a teeth-sucking sound and approached Mr. Jeong, the department head, a kind and soft-spoken man who always appeared temporarily sober. They whispered a few words, and Mr. Jeong looked at me and smiled and nodded. Mr. Lee came back and said it was okay to work from home.

I drove to my apartment, and in the parking lot was a Hyundai Sonata with its headlights on and doors open. A female soothsayer dressed in white danced and chanted while another woman banged a drum. Before the car was the severed head of a grinning pig with 10,000 won bills stuffed into its mouth. Incense wafted in the air. Plates of steamed pork sat on a bamboo mat, and a young man bowed before the pig while an older man splashed soju on the car tires.

I watched the car-blessing ceremony and went in and packed my bags with clothes and assorted gear. I grabbed my flip-phone charger, a box of books, and my file of Korean maps. In ten minutes, I was driving to the bridge that would take me to the mainland.

Nine hours later, I was approaching the Korean DMZ in the Gangwon Province. This is how you know you are getting close: Walls topped with razor wire run along the road, and armed soldiers stand watch at the entrance to the encampments. Army jeeps and trucks populate the roads and are driven by young men in fatigues. Military service is mandatory for men in South Korea, and it is a rite of passage for some and an eighteen-month prison sentence for others. Sons of the wealthy have been known to weasel out of this with fake medical exemptions. Other young men with special-ized skills could fulfill their service with a civil post, like my friend Dr. Lee Su-sup, who worked at a small rural clinic near Tongyeong in lieu of a military hitch.

Anti-tank barriers appear, and some are giant blocks of camo-painted concrete positioned over the road in a bridge-like formation. Motorists drive right under them. During an attack, the giant block drops down to obstruct the road. Others are along the edge of the road and take the form of cement blocks or huge boulders on pedestals. A closer look reveals detonation wires, and the charges are placed so the blocks and boulders tumble off the pedestals when blown and fall into the roadway.

Over the course of the next three days, I explored the area that Uncle Clayton had mentioned in the letter he had sent me over two years before. I visited the Punchbowl, based out of the town of Yanggu, a highland basin

Fig. 1. Anti-tank barriers near the DMZ are meant to be blown into the road in the event of a North Korean invasion. Photo by James Card.

surrounded by mountains. In the middle of it is the small town of Haean-myeon. I brought a small collection of books about the Korean War and read through passages referencing the Punchbowl. I found a few scattered monuments and historical markers, but trying to recreate how the battle unfolded was impossible without topographical maps and detailed infantry-level field notes.

At the Eulji Observatory, situated about a mile south of the DMZ, I looked upon the distant peaks of North Korea. I expected to see some action—tanks rumbling around or troop movements—but all was quiet as if the mountains were vacant of human activity. I visited the Yanggu Unification Hall, a museum and learning center about life in North Korea and how hopefully one day the two Koreas will be united in prelapsarian brotherhood. From there I went to the Fourth Tunnel, one of four tunnels (that are known) that the North Koreans dug to infiltrate the south. It was discovered in

1990. It is about six feet high and six feet wide and was designed to rapidly flood the countryside with North Korean troops.

After the tourist stops, I drove the back roads looking for bridges that might have been built during the Korean War and could have Red Jensen's welded signature. In case I found it, I had sheets of mulberry paper and sticks of charcoal and crayons to make a rubbing of the welded surface.

I drove the low country in the high mountains and stopped at bridges over small rivulets, channelized waterways, and irrigation ditches near farm fields. All the bridges were made of cement. There was no metal to weld upon. I ventured farther into the backcountry, and the bridges were constructed of the same cement but I noticed the creeks were prettier. They were freestone streams loaded with boulders and riffles and small waterfalls. They were very different from the trout streams where I grew up, in the Driftless Area of Wisconsin. While I stood under one bridge, admiring a plunge pool, a fish jumped out of the water and snapped at a mayfly. It looked like a trout.

3   Fishing the Pusan Perimeter

I quit counting how many I caught after the first two hours, around a couple of dozen. I had arrived at the Nakdong River just before daylight and
made my way to the shoreline with a headlamp. On the very first cast, the
strike had been so violent and the fish so big that the line snapped after a
couple of seconds of struggle. I tied up another Texas rig and worked a Gary
Yamamoto Senko onto the hook. The pumpkin-colored plastic worm was
the favorite for the largemouth bass, and it was a beautiful thing to figure
that out so early in the day.

The river curved and the sandy bank was too steep to traverse without
sliding into the water. I took to the high ground to circle back to the river. It
was covered in tall grass, and I hiked carefully and then I found the foxhole.
A week before I had fished the same spot, taken the same detour, and fallen
into it. I wasn't hurt. I was more fearful of snapping my rod than breaking
a bone. The grass and sandy dirt had cushioned my tumble.

I kept coming across foxholes. If you spend enough time hiking in the
woods, you will come across holes in the ground. Most of these are from
windthrow trees uprooted by storms. Eventually the tree becomes a nurse
log and slowly decays and replenishes the soil. All that is left is the hole,
but these holes were different. There were no explanatory root balls and
deteriorating trunks. These holes were perfectly round and had leveled
bottoms. They were man-made. There was no other explanation.

The foxholes I kept finding were located where they made sense: on
the Pusan Perimeter, the main battlefront during the early days of the

Korean War. It just so happened that this hill country held some of the best largemouth bass fishing in the country. As I hunted for bass, I also found the foxholes. The Pusan Perimeter stretches from the outskirts of Jinju to the west, and over to Masan, and then north to Daegu, and then east to Pohang. The line forms a square-like shape and encompasses most of the Gyeongnam Province. It's a vast area and multiple battles raged simultaneously along this fighting line.

This entire area was an incredible largemouth bass fishery. The Nam River flowing from Jinju meets the Nakdong in Changnyeong County, and here the river makes a big loop that stretches four miles east to west and five miles north to south. This area was the site of the Battle of the Nakdong Bulge.

All of the bridges had been destroyed, and on August 5, 1950, North Korean soldiers stripped down, bundled their clothes and rifles atop their heads, and waded across the river. In Changnyeong, not far from where I was fishing, I found a bronze memorial sign written in English and Korean:

ABSTRACT OF THE ESTABLISHMENT

IN THOSE DAYS OF THE KOREAN WAR, THE UNPARALLELED TRAGIC INTERNECINE WARFARE, HERE, IN PAKJIN AREA, U.N. AND REPUBLIC OF KOREA FORCES RESCUED THE FATE OF KOREA FROM THE LAST EXTREMITY WITH A DEATH STRUGGLE OF "NECK OR NOTHING." ON AUGUST 5, 1950, WHEN THE 4TH DIVISION, NORTH KOREA ARMY OCCUPIED YEONGSAN AND MILYANG, AND ATTACKED REPEATEDLY FOR ADVANCE TO PUSAN AFTER CROSSING THE NAKDONG RIVER, OUR REPUBLIC OF KOREA FORCES AND WARRIORS OF THE 24TH DIVISION, U.S. ARMY FRUSTRATED THE ATTACKED WITH PAINSTAKING COUNTERATTACK, AND FROM AUGUST 31 TO SEPTEMBER 15, THE 9TH DIVISION, NORTH KOREA ARMY ATTACKED DESPERATELY BUT OUR DEVELOPED AN ADVANCE AND RETREAT BLOODY FIGHT AND AT THE LAST STROKE A FATAL BLOW THEM AND DEFENDED TO DEATH A POSITION, AND THE FACT THAT U.N. FORCES COULD GET READY FOR A STEPPING STONE FOR ALL-

OUT COUNTERATTACK IS IMMORTALIZED IN HISTORY.
THIS HILL, WHERE A BEND OF THE NAKDONG RIVER
WINDS AROUND IS THE PLACE WHERE MANY PATRIOTS OF
OLD PASSED AWAY FOR OUR COUNTRY, SO WE ESTABLISH
THIS MONUMENT FOR PRAISING THEIR PATRIOTISM
AND INSPIRING DESCENDANT ANTI-COMMUNISM AND
THE SPIRIT OF DEFENSE OF THE FATHERLAND.

NOVEMBER 27, 1987

After exploring the rugged mountains near the DMZ in the Gangwon Province, I moved to the mainland. The English program at the shipyard was in limbo because of corporate restructuring, so I secured a new teaching job at the South Korean Naval Administration School across the water, on the mainland, in the seaside town of Jinhae. My girlfriend and I bought a car together (a beater Daewoo LeMans), and she visited me on weekends by taking the Guyeong Ferry across the strait.

With a base camp on the mainland, I could better access the *geumsu-gangsan*, the "land of embroidered rivers and mountains"—an ancient Chinese sobriquet for Korea. The problem with living on Koje Island is that it was expensive to get off the island. It cost time by having to drive across the bridge to Tongyeong and then hook up onto the expressway between the cities of Jinju and Masan. Often the traffic was savage and congested. Or it cost money by taking the car ferry to the mainland, and there was a chance the ferry might be filled up when you rolled in.

I now spearfished only during my visits back to Koje Island. The harbors and coastline of these seaside cities were murky and polluted, and summer red tides of decomposing algae clouded the waters of Jinhae and Masan Bays. Based in Jinhae, it was a quick drive to a network of expressways that could slingshot me anywhere in the country. I immediately started exploring the local waters and quickly started catching bass.

I knew that largemouth bass were present in Korean inland waters and it was an invasive species. It wasn't until I moved onto the mainland and had time to explore and fish that I realized how widespread they were. Largemouth bass were imported from the United States in 1972 when President Park Chung-hee devised the idea of commercial bass farms to add more protein to the Korean diet. In 1973 Togyo Reservoir, near the DMZ, was stocked with 500 bass. The National Marine Fisheries Service stocked 18,720 bass between 1976 and 1978. They were stocked in commercial fish farms in the Gyeonggi Province, which surrounds Seoul. Annual floods distributed the bass farther into inland waters.

The bass spread further than that, but the mystery was how they managed to migrate across mountains and into entirely new watersheds throughout the peninsula. This was done by a man called Mr. S., the owner of a tackle shop. He knew early on that largemouth bass could be Korea's top freshwater sportfish. In the mid-1980s, Mr. S. drove around with a fish tank built into his car and stocked lakes throughout the country. In America, introducing a fish species into a body of water where it doesn't belong is called "bucket biology," and it has messed up many freshwater ecosystems.

As more Koreans got into the sport of bass fishing, more anglers started transplanting their favorite new game fish. The same phenomenon happened in Japan, where such transplanters were called "bass guerrillas." Buddhist monks were another means of distribution. They performed the *bangsaeng* ritual, a ceremony that symbolizes love for living creatures, by releasing captive fish into rivers. They used bass for the rite, and little did they know they were unleashing a freshwater alpha predator into the local food chain. One story says that the wife of President Park, a devout Buddhist, personally introduced the species into Paldang Lake in this manner. By the 1980s there were concerns that the introduction of non-native bass wasn't such a good idea.

Along with largemouth bass, American bluegills were imported (*Lepomis macrochirus*) for fish farming schemes and, like the bass, they escaped and formed fugitive populations in watersheds. Their populations were more localized and harder to find as they are a less popular sport fish than the largemouth bass and nobody was actively transplanting them around the country.

Environmentalists claimed the bass were eating everything—which, of course, bass naturally do. The Ministry of the Environment added bass to a list of non-native predators and mentioned the fish in some public awareness campaigns about invasive species.

Korea's carp fishermen complained that largemouth bass were decimating their beloved stocks of crucian carp (*Carassius carassius*). Korean

Fig. 2. Invasive non-native largemouth bass, *Micropterus salmoides*, are widespread through-out the Korean peninsula. Photo by James Card.

anglers love to pursue this fish, and it has been stocked widely, dumped in reservoirs throughout the country. The only thing that impressed me about fishing for crucian carp were the bobbers that were used. This bobber was a long wooden needle with a bulge in the middle. Some were over a foot long, and they were as elegant as fishing tackle and deadly accurate at detecting the lightest bite.

As for bluegill, carp fishermen tossed them onshore to rot. Common sights around carp angling points are the dried husks of bass and bluegill discarded by embittered carp fishermen. They look upon both species with contempt, and I often came across fishing spots and found their skeletal remains. This was very absurd to me: one of the world's most delicious panfish was being thrown aside for carp.

Complain as they might about bass, the carp fishermen did not seem to consider overfishing in their ranks as a cause of decreased fish stocks. There are no fishing licenses in South Korea and very few angling regulations. There are no game wardens. Anglers can keep an unlimited number of any fish of any size. Besides rods, they use cast nets, gill nets, trammel nets, trot lines, and jug floats. Fisheries conservation is not a varsity sport in South Korea.

However, a minority of Korean fishermen have come to appreciate the sporting value of pursuing the largemouth bass. It is easy to understand the appeal: since the early '70s, the largemouth bass has been spreading and increasing in numbers without heavy fishing pressure other than accidental and occasional bycatch by the carp anglers. It takes about ten years for a bass in temperate climates to reach ten pounds, but in ideal conditions the fish can grow faster and larger. For decades, the largemouth bass have been getting bigger and more numerous while growing relatively unmolested. Other than a few thousand bass anglers in a country of fifty million people, it's a mostly untouched sport fishery.

Tackle shops cater to their bass angling customers, offering assorted bass lures. Websites and clubs have been formed, along with the Korean Bass Federation, which holds bass fishing tournaments. Largemouths are found in the Han River system in the Seoul area and its Kyeonggi-do tributaries. The Nam River connects to the Nakdong River, and bass can be found in points along both river systems. The Guemho River, which wraps around the north of Daegu, gets a lot of action with local bass anglers. Andong Lake is a popular location for bass, and tournaments have been held there. Connected to these rivers are canals and creeks that connect to hundreds of other sloughs and reservoirs.

It is possible to land a large bass that has never had a lure cast to it. One reason is the lack of personal watercraft on Korean inland waters. Ranger-style bass boats are few and are seen only on the mega-impoundments—imported by Korean bass aficionados at enormous expense. Rubber rafts with trolling motors are spotted only here and there, and fishing from canoes and kayaks is almost nonexistent. This leaves most anglers fishing from shore on well-worn paths that lead to well-pounded spots, and the plastic worms hanging in the nearby willow branches attest to it. What remains is the untouchable

Fig. 3. Invasive non-native bluegill, *Lepomis macrochirus*, are overlooked as a game fish and grow huge in hard-to-access spots. Photo by James Card.

and inaccessible water. For the anglers that use waders, a belly boat, or an inflatable pontoon, vast untapped water opens up.

This created an odd duality. Many of these bass anglers have adopted the conservation philosophy of catch-and-release fishing. Returning the fish to the water ensures the fish will grow larger and continue to spawn. Ironically, by practicing catch and release, the anglers are perpetuating the problem of the largemouth bass being an invasive predator that out-competes the native fish.

There is no easy answer, especially when considering their recreational value—and also the simple fact that they are here to stay and have now thoroughly permeated some Korean inland waters. There is no way to eradicate them completely. To prevent further distribution, it is illegal to transport the fish and place it in other bodies of water.

This is what I pondered when I stopped for brunch. I decided to create my own policy for dealing with the bass I caught. The Nakdong is South Korea's longest river, and keeping or throwing back bass would not make a difference. It also runs through urban and agricultural areas before it empties into an estuary outside of Busan. It isn't the cleanest water to consume fish from, and it was the site of South Korea's biggest industrial pollution case. In 1991 thirty metric tons of phenol leaked into the river from the Doosan Electronics plant near the city of Gumi, and it polluted the drinking water of Daegu, the country's third largest city. Thousands of people became violently ill. A month later, 1.3 tons of phenol leaked into the river.

The Nakdong also runs through a lot of agricultural fields, and I often wondered about runoff. In my explorations in this region, I came across many small reservoirs, nestled up in hillside valleys, that were stocked with bass. They were surrounded by forest, and the water was relatively clean and clear. So this was the rule: throw back the big-river bass and keep the small-reservoir bass for eating.

I worked along the shoreline, my boots clumped in mud. Around a bend, a gray heron waited downstream for small fish to pass within striking distance. *Ardea cinerea* is a consummate solitary hunter, and I've seen them swallow largemouth bass whole and headfirst. The heronry of the Nakdong watershed is home to many of them, and their nesting platforms, in old pine trees, are up and down the river. A month earlier I had witnessed their courtship dance. The male stretched out and flared his neck and clappered his yellow bill. He bowed to his beau and she bowed back, and they skipped to and fro with open wings.

At an ancient petrified log, barely visible under the surface, I felt a hard tug and the bass shot out of the river and tail-walked on the water. That was enough for the great bird. It stretched its wings—spanning six feet— and rose into the air with slow wing beats. It coasted farther downstream, its pterodactyl-like voice croaking out a guttural "Fraaaawk, fraaaawk." I felt bad for disturbing the fellow fish hunter. The bass could have been his lunch. I brought the fish to net—a nice twenty-incher—and released it back into the river.

I had a can of Spam tucked in my backpack. I kept a box of them in the LeMans as road food/survival food. I did not buy the Spam. They were gifts from Korean co-teachers and friends. South Korea has an odd appreciation for Hormel's classic meat product, which dates to the Korean War, when American GIs introduced it to the locals. Now they buy millions of dollars worth of Spam every year, and it has become a treasured holiday gift. The

Spam given to me came in fancy boxed sets holding a dozen cans. The boxed sets piled up until I stashed them in the LeMans. Out of the tin, I found the pink, gelatinous meat unappetizing, so I made a small driftwood fire on the sandbar and I whittled a green willow stick to a point. I diced the Spam into chunks and roasted the mystery meat until the grease dripped into the flames and the outside was crispy and sizzling.

I wasn't the first guy to eat a can of Spam along the Nakdong River. From August to September in 1950, thousands of American soldiers fought in these hills and they had to eat something, despite nicknaming it "the ham that didn't pass its physical." I thought of how many empty tins of Spam had been littered across the landscape—along with shell casings and blown-up machines of war and abandoned weapons and tools. I always kept an eye out for such an artifact but never came across anything. I daydreamed of finding an M1 Garand dropped and lost between a couple of rocks. I surmised that Koreans in the postwar years scavenged every scrap of metal for trade. I only came across the foxholes.

I washed out the taste of Spam in my mouth with what I called Korean Gatorade. In my visits to many tea houses, I had settled on one kind of tea that I liked. I brewed it at home, cooled it in the fridge in a glass pitcher, and filled my water bottle with it before fishing. It was *omija*: the "o" is five, the "mi" is taste, and the "ja" is tea. It's a tea of five tastes: sour, sweet, spicy, bitter, and salty. The tea is made from the red berries of a woody deciduous vine (*Schisandra chinensis*) found in Korea, Russia, and northern China. Like most things in Asian herbal medicine, it supposedly can cure or prevent every malady known to mankind. It also enhances mental and physical performance and doubles as an aphrodisiac.

I looked out at the river. Any soldier tuning in to the radio during the summer of 1950 would have heard the voice of Seoul City Sue, Anna Wallace Suhr, an American Methodist missionary who married a Korean man with left-leaning views. When the Korean War broke out, they ended up on the North Korean side, and she was tasked with reading propaganda and the names of dead American soldiers between aired Sousa marches. There are multiple accounts of the Nakdong River running red with blood and bodies floating down the river. On the sandbar the river was about knee deep, and then it dropped off to a depth of about nine or ten feet. The invading North Koreans built rafts to cross the river. On August 27th, 1950, four men of K

Company of the Thirty-Eighth Infantry volunteered to destroy the rafts. They floated half-filled drums of gas across the river. Under heavy enemy fire, they soaked the rafts with gasoline and torched them. They made it back to shore, and later all of them were awarded the Silver Star.

North Korean soldiers also built underwater bridges with logs, sandbags, and rocks. They kept throwing shit into the river until they could ford across. With the water muddied by the war activity, they were hard for pilots to spot, but an angler would be able to detect them. There would be riffles and unnatural flows, and with the new mass in the current, the water would push past the natural high-water mark. I wondered if some of the logs were still there, buried in silt and providing cover for bass.

I fished for bass less and less. My freezer was full of bass and bluegill fillets. The initial excitement of catching America's greatest game fish halfway around the world was wearing off. I grew up catching largemouth bass, and I loved the fish but only to an extent. I grew up reading the Bass Pro Shops mail-order catalogs, but later I found the growing commercialization of the species through television shows and tournaments to be crass and lame. After a while I quit fishing for bass on the Nakdong and only focused on some backcountry reservoirs and smaller tributaries and fished for the better-tasting bluegills. The mission was simple: catch and kill a mess of bluegill to fillet and freeze for a Friday night fish fry with friends and no more. It was a turning point where I slowly learned something about myself: I became more excited when the fish I caught wasn't a bass or a bluegill. It did not matter if it was big or small just as long as it was some other species.

While throwing Mepps spinners and Rapalas for bass, I sometimes caught an odd silvery fish that I could not identify. It was aggressive to the lure and fought hard. Curious, I later identified it in a fish guidebook and asked other anglers about it. In Korean, it has more than ten different names in regional dialects, but most call it *kkueri* (pronounced with a hard k/g sound). Its scientific name is *Opsariichthys uncirostris amurensis*, with the last word denoting that it is native to the Amur River basin. It is also native to Korea, eastern Siberia, Japan, and China. The Chinese call it the "horse mouth." Among Korean anglers, they aren't considered a good-eating fish, and they call them *sonnim gogi*, or "guest fish." As in, they are not the primary fish of pursuit, more like a passing-through bycatch.

In English, it's referred to as the Korean piscivorous chub, with *piscivorous*

Fig. 4. The Korean piscivorous chub ("notchmouth"), *Opsariichthys uncirostris amurensis*, is a fish-hunting carp that is excellent on light tackle. Photo by James Card.

meaning it is a fish eater. I nicknamed it the "notchmouth." Its mouth has a notched upper jaw that fits neatly into a hooked lower jaw. It is this identifying characteristic that reveals its true nature, that it is a hunter of other fish. It is the only member of the omnivorous carp family that has evolved into a flesh-eating predator.

I noticed they hit my lure only near the surface and during fast retrieves. I swapped out my spinning gear and started fly fishing for them. I threw out Mickey Finns, Hornbergs, woolly buggers, muddler minnows, and gray ghosts tied with a perfection loop for freer movement. All of them worked in any color, but I kept the size of the streamers small. The average size of notchmouths is between eight and twelve inches, and the big ones measure at around seventeen inches. They have a brownish-green back and silvery white belly with the kind of scales one would find on a member of the carp family.

The trick is a fast retrieve. You want to strip as much line in as fast as possible. Notchmouths are killers of moving targets, and often they miss the streamer, sometimes smacking it but not getting hooked. It is almost like the bump-and-bite behavior of sharks attacking their prey. Perhaps it is part of their hunting behavior. It happens enough to be observable. They also jump out of the water in a slashing leap for topwater flies, and I chugged small poppers and skittered foam spiders across the surface to elicit an attack.

It was while fly fishing the area of the Nakdong Bulge for notchmouths that I made two first discoveries. As I was wading through some tall grass and looking out for foxholes, I flushed a Korean water deer (*Hydropotes inermis argyropus*). It was about the size of a whitetail fawn after it loses its spots. The deer bolted along the edge of a rice paddy and dove into the thick cover of an irrigation ditch.

Hydropotes refers to the Greek prefix *hydro*, meaning water, and the Greek *potes*, meaning drinker. *Inermis* in Latin means "unarmed," and it is one of only two living species of deer that do not possess antlers. There is another subspecies in China. The *argyropus* subspecies is native to the Korean Peninsula. While not possessing antlers, the water deer has canine-like tusks that protrude from its upper jaw. Both sexes have tusks, with the buck's being considerably larger.

I had seen a mounted specimen at a museum on Koje Island. The taxidermy was hideous and deranged but decent enough for me to get a sense of the animal. The tusks are more like fangs, and they protrude from their mouths enough to be noticeable. I'd seen their tracks countless times in the mud and sand along the riverbanks, and while studying a set of tracks, I looked upriver to see where they had gone and I found my first bryozoan.

It looked like a zombie brain impaled by a driftwood stick. They are ancient aquatic invertebrates and filter plankton from the water. The bryozoan resembles a brown, gelatinous blob and often clings to a structure like a tree branch. The rosettes on the surface—which lend it a mutated brain-like appearance—are individuals that form the volleyball-sized colony. They are harmless and the only problems they might cause are plugging inflow or outflow pipes. In autumn they break apart and disintegrate as part of their life cycle.

Fishing among old battlegrounds.
An untapped largemouth bass fishery.
Giant slab bluegills never before hooked.
Hyperaggressive predatory carp.
Vampire-fanged deer.
Herons like winged dinosaurs.
Aquatic zombie brains.

I wanted more.

4 The Boy at Toad-Swarm River

I learned about *sogari* through Mr. Na. A mutual acquaintance introduced him to me, and we agreed to meet at a rest area near Sancheong at 4:00 a.m. on a Sunday morning. I parked my vehicle there and rode with Mr. Na to the Gyeongho River in his SsangYong Musso, a tank-like jeep. He was an engineer at a small shipyard, and his wife was an airline attendant for Asiana. I told him he could call me Jim; it was my nickname. He told me his nickname was *gae-dwaeji*. Dog-pig.

"Because I eat like a pig and my face looks like a dog," he said.

I looked at him more closely. He certainly had a bulldog face, and his build was stocky. He had thick wrists, and he had hairier arms than most Korean men.

I asked about his wife. Asian airlines are famous for discriminatory hiring practices, having height, weight, and beauty requirements for their stewardesses. He showed me a picture of her, and she was stunning, with seductive eyes. I asked where she usually flew. All over, he said, and rattled off Sydney, Hong Kong, Jakarta, Osaka, Saipan. She wasn't home much. They did not have any children. He grew quiet and we didn't speak until we got to the river.

One of the most gorgeous women in South Korea was married to a man named Dog-pig. I looked at him again when we got out of his Musso. I could see how his wife had fallen for him. He looked strong, like a heavyweight boxer, the kind of man who could protect a beautiful woman. If he were to play a role in a movie, he would be the strongman Korean that nobody wants to fuck with. He would be a hero, but in modern Korea he was an antihero, a living anachronism. There was a trend in South Korea in which femininity in men was celebrated. Men were getting plastic surgery and wearing makeup. Celebrity K-pop boy bands exuded more estrogen than teenage girls. It was a metrosexual look of pale, smooth, hairless skin, girly haircuts, soft voices, and a persona that oozed weakness. Mr. Na's toughness ran in the family as he told me that his cousin was one of the Roof Koreans who armed themselves to protect their property during the 1992 Los Angeles riots.

There was a melancholy about Mr. Na, and I believed it was from his wife being away. Not just any wife but a supermodel of a wife who was flying to exotic destinations and staying alone at hotels in big cities and coming in contact with hundreds of traveling men. A husband's imagination can lead to dark thoughts. When his wife was traveling, Mr. Na put all his energy into fishing for sogari. He lived on what I called "super ramyeon." On his butane camp stove, he boiled a pot of water and dumped in one pack of ramyeon, one can of tuna, and a couple of eggs. He would spend nights out there alone, sleeping hours at a time in the back seat of his Musso.

Instead of sleeping with his wife, he would sleep with smooth stones he found in the river. They were called *suseok*, or scholar's stones. They were odd-shaped rocks with abstract beauty—some of them resembled mountains. One he showed me was a small clam-shell fossil; another stone had white quartzite spiderweb striations. He piled them in the footwell of the back seat.

We were after a fish that artists of long ago used to symbolize the Korean king. Only one fish was depicted in a painting, never two; since there was only one king to be loyal to, the fish swam alone on the canvas. It is called the mandarin fish in English and sogari in Korean. The scales of *Siniperca scherzeri* are a collage of golden-brown hues forming leopard-like rosette patterns. The average catch is eight to twelve inches in length, with a specimen over twenty-four inches regarded as a rare trophy. They do not get a chance to grow very big as they are coveted as a delicacy, eaten raw, steamed, or in a hot peppery soup called *sogari maeuntang*. I knew largemouth bass anglers who practiced catch and release, but Mr. Na was the first Korean angler I had met who did so with native species. He said he had kept a few in the past but preferred now to let them go.

Fig. 5. The Mandarin fish, *Siniperca scherzeri*, is a prized game fish among Korean anglers. Photo by James Card.

The mandarin fish has a lower jaw longer than the upper jaw. It is in the perch family and, like the yellow perch of North America, they like to hold tightly to the bottom. But they are also similar to smallmouth bass in that they are found in fast-flowing, cool-water rivers. Mr. Na had found a source for unpainted lead jigheads, and he bought them by the pound, along with soft plastic curly-tail grubs bought in bulk.

I have heard of Korean largemouth bass anglers occasionally picking up a mandarin fish on a crankbait or some other lure, but Mr. Na was adamant about using only weighted jigs. He believed it was the only lure that consistently worked. He was right and he out-fished me every time we fished the Gyeongho together, often catching ten to my one. It came at a cost, and that was losing the weighted jigs on snags. The boulders and rocks in the Gyeongho were mostly smooth, so instead of snagging on the rocks, the head probably got wedged in rock crevices. Mr. Na wanted it that way. That meant he was fishing correctly. He believed the best presentation was to very slowly hump and wiggle that tiny jig over, under, and along every nook and bump along the river bottom. Retying a jig that had broken off ten times in one hour was common.

That was frustrating and wasteful, but I also found it very boring. What we were doing was like Czech nymphing, which is a fly-fishing technique using a weighted fly to drift near the bottom. I missed the topwater action and, to make it worse, we often fished in the dark with headlamps, and I often had no idea how my jig was behaving underwater other than by the feel of the line and rod tip.

One morning Mr. Na obsessively worked over a deep-water pool, presenting the jig over every inch of the river bottom. I slipped to the end of the pool, where the water flattened out into some shallow riffles. Sogari are in deeper water, so I quickly reeled in my line to skip ahead downstream. A fish smashed the jig on the retrieve. I yelled and got Mr. Na's attention. He walked over and shrugged.

"It's a *keokji*," he said, nonplussed. I brought in the small perch-like fish. Its body was a mottled camouflaged color with dark tiger stripes, and it had a dark spot behind each gill plate like a bluegill. "They are too easy to catch," he said and walked back to the pool.

That moment was the end of fishing with Mr. Na. I was tired of his tedious nocturnal scouring of the river bottom for a very moody fish, and I started exclusively fishing for keokjis. He looked down on this. He came to pursue mandarins, the fish of kings, not this small species, and we eventually drifted apart over this preference. Our phone calls to go fishing with each other dried up.

I fished the Gyeongho on my own and targeted keokjis, also called the Korean brook perch and the aucha perch—although I never learned where the word *aucha* came from. While the mandarin fish is distributed throughout Manchurian China, the Russian Far East, and the Korean Peninsula, *Coreoperca herzi* is a true native of Korea, and for me that was an enormous attraction—that this small, hard-fighting fish existed only in this part of the world, and in this country, it existed only in cool-and cold-water streams. It was a rare and savage species of fish.

They do not get very large—a ten- or eleven-inch fish would be a trophy— but they punch far above their weight class. They have the aggressiveness of American smallmouth bass, and they like to hunt in shallow rapids that are

Fig. 6. Korean brook perch, *Coreoperca herzi*, are native only to the rivers of South Korea and fight in a way that is similar to smallmouth bass. Photo by James Card.

a mix of pebbles, small boulders, and river cobble. Like the notchmouths, I gave them a more colloquial nickname: smashmouth perch. They were a perfect game fish for a four-weight fly rod, and I cast all my small streamer patterns at them and they ravaged them on a quick-strip retrieve. To extend my reach, I waded for them and I would sometimes see them shadow-darting along the river bottom. They were everywhere.

I married the elementary school teacher. We moved to Gwangyang, farther west along the southern coast. She transferred to a one-room schoolhouse, where she taught multiple grades in the same room. Children in rural areas were so rare that grades and classrooms were doubled up and consolidated. Many rural elementary schools were closed for a lack of students. I once came across an abandoned school near the Nakdong River, and bamboo and pine were already forming thickets around it. On that dead-still day, with no wind and no breeze, one swing swayed back and forth at the playground.

I got a new job with POSCO, the giant steelmaker. The organization was so big they funded their own private school system for their employees' children. Our workplaces illustrated the dichotomy of South Korea's end game: you either work for a *chaebol* or you don't. If you work for a chaebol, your children will go to the better schools and have an all-around better life. If you don't work for a chaebol, you might work for the government, or you are destined to work at a lower-tier company or be forced into entrepreneurship, for better or worse.

The chaebol are the family-owned conglomerates that dominate the lives of all South Koreans. The four super-chaebols of South Korea are Hyundai, Samsung, Daewoo, and Lucky Goldstar. They dominate the national economy and have assets in the billions. There are smaller chaebols like Woori (banking), Haitai (foods), and Shinsegae (department stores). Not only do they dominate their primary industries but, through their holding companies, they deal in other fields, including hotels, insurance, and petrochemicals. Every aspect of life is dominated by the chaebols. You are born in a chaebol-owned hospital, you brush your teeth with chaebol-produced toothpaste, you live in a chaebol-constructed apartment, and when you die your spouse is paid off with a chaebol-underwritten life insurance policy.

President Park Chung-hee believed steel was the backbone of a nation's industry. He nationalized the banks and encouraged the chaebol model of business growth. He enlisted Park Tae-joon to develop an empire of steel. POSCO started as the state-owned Pohang Iron and Steel Company in 1968. Their first steel mill was built in Pohang, and they expanded to the small fishing village of Gwangyang in 1987. Decades later, POSCO became one of the world's largest steelmakers and one of the world's biggest corporations.

The world of POSCO was more communist than capitalist, and it was evident in the central planning. We lived in Geumho-dong, a neighborhood built on a reclaimed island. Only POSCO people lived there. Unlike the rest of the country, the streets were wide and all the utility wires were buried underground. There was none of the gaudy neon or other ostentatious signage found in all other Korean communities. There were parks, fields, well-tended topiary, and wooded hills of pine and rhododendron. The housing was slightly above average compared to the rest of the country. There were strip malls that catered to everyone's basic needs. The island had the vibe of a proletarian utopia designed by a benevolent neo-Confucian employer.

It was a great place to work and raise a family but very dull. As newlywed DINKS (dual income, no kids), we jetted off to India, Vietnam, Thailand, and Cambodia during our school vacations. The coastal waters around Gwangyang were polluted and murky from the massive-scale industry, so on weekends we cruised over to Namhae Island to spearfish and hang out at the beaches. It is South Korea's fifth largest island, and it reminded my wife of her home island of Koje but without the shipyard culture and economy. It is where Admiral Yi Sun-shin died in battle after taking a bullet from an arquebus.

I hiked and explored Mangwoon Mountain, the crash site of a U.S. Army Air Corps B-24 "Lady Luck II" bomber. These aircraft were painted black, and they were nicknamed "Snoopers," as they flew night missions to bomb enemy shipping at low altitudes. On August 7, 1945, while flying over Namhae Island, the bomber smashed into the 3,000-foot peak, and all eleven airmen perished. They were the only U.S. casualties in Korea during World War II. Korea was occupied by the Japanese, and authorities scavenged the wreckage and left the burned bodies of the airmen to rot. A local man, Kim Duk-hyung, gave the airmen a proper burial and marked the spot with a pinewood cross. His brother had died in a plane crash in Burma and was never found. For this act of decency, the Japanese tortured and imprisoned him. After the war he was a free man, and he later led a U.S. army captain to the gravesite so the bodies could be recovered. Kim built a memorial at the location, and the U.S. military has honored him with awards and citations.

All of my old fishing spots in the Gyeongnam Province were farther away. They were too distant for quick afterschool run-and-gun angling outings. I needed to find some local places to fish. When samurai warlord Hideyoshi invaded Korea in the 1590s, millions of toads swarmed the Seomjin River, and their ghastly croaking prevented the Japanese army from crossing. The river rises in Maisan (a mountain with twin peaks that look like horse ears) and flows 131 miles into Gwangyang Bay, not far from my new apartment.

The Seomjin is the fourth largest river in South Korea, and it is very clean once you get past the industrial area near Gwangyang. The river did not flow through any industrial areas, only through small farmlands bordering the foothills of the Jiri Mountains. I also discovered there were no largemouth bass in the river, only native fish. As you get on Highway 19 outside of Gwangyang, the road winds up a hillside, and you can look down and see old women rake the sand flats in the brackish zone for *jae-chop*, corbicula clams. They fill their woven baskets with the tiny shellfish and carry them off on their heads. It is an old tradition to use them in a spring-onion broth. Around this estuarine zone I would sometimes spot common shelducks and black brants flocked on the flats.

It's a part of the country where the flowers bloom earlier in the year. The flowers of the small green *maesil* plum trees mark the start of spring. They are one of the first to bloom, a few weeks earlier than cherry blossoms or yellow dogwood. The typical budding season is from February to late March. These blooming trees line the road, and in the early spring it is a colorful alameda. The Seomjin is a sandy river, and the town of Hadong was once called "the county of much sand." The region was known for its porcelain kilns, and today artisans make stoneware tea bowls with the local clay. Deeper into the valley barley fields replace rice fields.

I spent many afternoons in the Seomjin River valley fishing for keokji and sogari. I fly fished more often and reached for my spinning tackle only when I came across a deep hole where a sogari might lurk. There I used the inch-by-inch dredging technique Mr. Na taught me. Along a trail I used to cut back to the river was a mushroom nursery. A long bamboo pole was mounted at chest height between the shade of two chestnut trees. Alternating logs of Mongolian oak leaned against the bamboo crossbeam, and *pyo-go* mushrooms sprouted from the logs. It is also known, by its Japanese name, as the shitake mushroom (*Lentinus edodes*), and the dark brown caps have a meaty, chewy texture. I watched the old farmer inoculate the logs by hammering a mycelium-covered plug into the many holes that he had drilled.

On every trip I would drive deeper into the valley. I would have vast stretches of river to myself. The only time I saw other anglers was in the autumn during the shad run. On the lower stretch of the Seomjin, near where the freshwater mingles with saltwater, I found anglers fishing for *jeoneo*, known as spotted gizzard shad. It is like the American gizzard shad and a great eating fish. There is an old saying that the smell of gizzard shad on the grill is so appetizing that it will bring back runaway daughters-in-law. I read John McPhee's book *The Founding Fish* to learn more about the species. But the most interesting thing I learned was how some Korean anglers fished for them.

Some used nets and other contraptions to catch the shad as there are no fishing regulations in South Korea. No fishing licenses, no bag limits, no rules. The only restriction I came across was a No Fishing sign posted at a reservoir used for drinking water; the operators did not want slob fishermen mucking the place up.

There were two fly angling techniques that I found to be of interest. One was a cross between Japanese tenkara fishing and spey casting. A tenka rod is a telescoping rod with no reel. At the very tip is a nub of cord that the leader is tied onto. This rig was developed to fly fish freestone streams in the mountains by dapping and flicking the fly into riffles and runs. Koreans have a similar rod; it is longer and is more often utilized as a very long cane pole. Many anglers use them rigged with ultra-thin pencil bobbers to detect the delicate bites of crucian carp, panfish-sized carp that are beloved among many Korean freshwater anglers. When hooked into one, the angler merely lifts the long rod and pendulum swings the fish toward him.

The angler on the Seomjin had the same kind of reel-less cord-tipped rod but had it rigged with a long leader with a tippet of multiple dropper flies. The flies were generic-looking wet flies—swept-wing and fuzzy and buggy. He drew the rod back doublehanded across his shoulder, as one would do a roll cast, and then slung the line downstream on the forward cast. He let the line stretch out in the current and then gently jigged the line upstream and let it drift back. The technique worked and he caught shad.

The other angler used a *gyeongi*; it was a short rod, about two feet long, topped with a helix-shaped ladder. The fishing line was wrapped around the helix, and the line was extended and brought in by spinning the shaft of the rod in a finger motion like rolling a joint. The flies were the same as the other angler's, and with this rig, all one can do is let out the line into the current and slowly work it back in. It was a primitive device and could really only be used for catching small fish.

Other than the shad anglers, I mostly had the Seomjin to myself during weekdays. On weekends there would be a few out casting around, along with more sightseeing traffic driving up the river valley. The day I met the boy I was fishing a stretch south of the town of Gurye.

I had fished the area the night before until dark, and while hiking back with my flashlight I spotted a Eurasian otter while scanning the shoreline. Its eyes glowed in the beam, and it swam closer, drawn to the light, close enough to see its wet whiskers. It was a good sign for the health of the Seomjin as an otter would not live here if the fishing were poor. While shortcutting my way back to the car, I bumped a water deer, and it crashed through the brush and made a warning bark that sounded like a shrieking growl.

The next day I picked up where I had left off, and I found the otter's five-toed paw prints and the fishbone remains of a mandarin fish. On a small knoll, I found an oval of matted grass where the water deer had bedded down. Among the grass were white hairs from its underbelly. Nearby small saplings were frayed and gnawed by its tusks, and the buds on the yellow dogwood had been browsed off. When I mentioned my sightings of this deer to a Korean colleague, he said there was an old superstition that the bite of the deer is fatal. It might not kill you, but it would be like being gouged by a small pocketknife. Both sexes have canine-like tusks, with the buck's being larger—up to three inches in length and shaped like a curved dagger. During the rut, it thrusts its tusks and gores the opponent's flank and neck.

I cast out a muddler minnow for the smashmouth perch and worked my way upstream and let the current grab the line downstream and straighten it out. Retrieve it in, take three steps upstream, and repeat. I felt a tug on the second cast, and the three-weight bent over and quivered. The brook perch fought back with a zigzag across the shallow river cobble. I brought it in, and it was about the size of my hand. I studied its camouflage and its huge mouth and understood how it had evolved to be an ambush predator hidden in the river stones. I released the little hunter and moved upriver.

As I stepped forward, I heard a crunching sound underfoot. The sound of crumpling plastic. The white plastic bottle had once held paraquat. I had found empty bottles of paraquat along the riverbanks before. Drinking this herbicide was the most common method of suicide for older rural folk. South Korea has one of the highest suicide rates in the world, and by occupation farmers make up the largest group.

South Korean agriculture was sacrificed to the country's vision of modernizing into an industrial—and later digital—player in the world economy. Some believe the underlying cause of the high suicide rate is the country's rapid transition from a cooperative agrarian culture to a wealthier urban rat race after the Korean War. The countryside was emptied for the bright lights of the big factories in the big cities, and only the old were left to tend to the family farms.

The most common farm machine is the two-wheeled tractor: the iron ox, the one-eyed buffalo. It has one axle and is self-propelled. It is extremely versatile and can pull a trailer and power other implements. I've never seen a man under fifty operate one. In the fields and orchards, grandmothers do the kind of work grandsons are supposed to do. There is no retirement, only survival.

The birthrate in South Korea is one of the lowest in the world, but it is even lower in the country because there are no young people. Many mountain villages have not heard the cry of a baby in years. Nine out of ten Koreans live in cities. These country villages are super-aging societies, but not long ago they were places of joyous family life, civic-minded spiritedness, and community-invested industry. In 1970 President Park Chung-hee launched the Saemaul (New Village) Movement. It was like a self-help program for rural areas that was funded by the government. With a cement surplus, he ordered that each village be awarded 335 bags of cement. Villages that used all of that were given more cement and steel bars. Roads were paved, rice paddy irrigation was reinforced, bridges were built, and town halls were erected.

While mulling over the empty paraquat bottle, I realized this was why I had all of these rivers to myself—not just the Seomjin, but nearly everywhere I went. And not just fishing, but hiking and exploring the backwoods and small hidden valleys. There were no young people and very few middle-aged people. Everyone else in the nation was working sixty hours a week in urban areas and was too stressed out to screw and make babies.

The underlying reason my angling adventures were peaceful and successful was because there was hardly anyone around to interrupt me. A person might think that to have a sweet fishing spot all to yourself you might have to go far off the beaten path to a remote area. That is true on the Korean Peninsula and universally true anywhere, but a strange collision of war, industrialization, suicide, stress, declining birth rates, and urban migration left the countryside emptied of much human activity other than old folks scraping by to make a living. The emptied countryside was like a dystopian film set where only a few survivors remain. This sounds grim and the suicide statistics do not lie, yet there was still much happiness in these mountain valleys. The old timers that I came across always welcomed me and were friendly and curious. They often offered food and rice wine. They were humble and polite. Few foreign visitors ever make it into these remote corners of the land, much less one with a fishing rod.

The boy stood on a one-lane Saemaul-era bridge and dangled a line off a bamboo pole into the plunge pool underneath. I watched him and could tell he did not know what he was doing. He was about twelve years old. I walked over to him and asked if he had caught anything. He said nothing and looked at my rod and vest. The concrete bridge had no side walls or curb, just a bare edge. I asked him where he lived, and he said with his grandparents and pointed to a small farmhouse tucked into a far hillside surrounded by *maehwa* plum trees.

I had heard of Korean parents sending disabled or pregnant kids back to live on the family farm. There weren't any support systems in the schools for them, and they had no chance of succeeding in this hypercompetitive society. The best they could do was let the kid live a good and simple life in the country surrounded by blood relatives. This boy looked bright and active, with no apparent disabilities, which meant he might have come from a broken home or maybe his parents were down and out and could not take care of him. Perhaps he was an orphan.

He was the first kid I had ever seen fishing in South Korea by himself. The boy lifted his line back in. On it was nothing but a bare hook, not even a sinker. The boy laid the pole down on the concrete and scampered off. He pushed over rocks looking for worms and bugs. I walked off and did not say goodbye or good luck. I hiked down the road that led to the bridge, slipped along some barley fields, walked the river's edge, and cut over to a tractor road that led a couple of miles back to my car. I stowed my gear and drove back to the boy.

He must have found a worm because his line was back in the water. I brought out my spinning rod with a lead-head jig tied on. "Here," I said in Korean. "Watch." I demonstrated how the bail opened that let out the line and how to pinch the line against the index finger to hold it and later release it at the right moment of the cast. I threw out some easy casts, some gentle lobs, and reeled them back in. He stood there like a good student, quiet and watching my every move. I made some longer and more precise casts to show him how effectively and wonderfully this tool worked, and then I handed him the rod.

I thought for a moment about the reel handle and looked at the boy's hands and remembered all Korean kids are right-handed. Lefties that do not use their right hands get their left-handedness smacked out of them with a ruler or a kitchen spoon. The boy awkwardly opened the bail but did not pinch the line, and the line spooled out. I told him to reel it back in. He did until the jig was buried in the rod-tip eyelet. I took the rod back from him, dropped the jig out, and reeled it back in until about six inches of line hung from the tip. I explained you wanted to leave a little hanging out on the end. I gave him back the rod, and he made his first cast.

We practiced casting for the next half hour and spoke few words. His English was nonexistent other than knowing some loan words, and my Korean was good enough to coach him on the basics. I said to him, "*Dahshi.*" Again. Every time he brought the lure back in: *dahshi.* Cast again, again, again, again, like a martial arts master commanding a student to build muscle memory in a movement. Each cast was better than the one before. I snipped off the jig and tied on a small Mepps Algia. I told him that fishing from the top of a bridge was a stupid place to fish. In my field notebook, I sketched a stick figure on a bridge holding a fishing rod, and down below I drew a Jesus fish. From the eyeball of the fish, I drew a vision cone radiating upward, and the cone covered the body of the angler figure.

I whispered, "They can see us. *Katchi Kapshida!*" (Let's go together!)

We snuck off the bridge, staying low like we were avoiding helicopter blades, and crawled down the rocky slope to the shoreline. We went downstream and commando crouched near a flat of riffled water near a large boulder. I told him to cast out and reel it in very fast: *bali, bali.* Hurry, hurry. He bombed a cast into the riffles and furiously cranked the reel and then grunted and moaned in confusion. The reel was harder to crank in. I watched the rod tip. He had a fish on and did not know it.

Halfway into the retrieve, the keokji splashed the surface with a muscular flutter, and the boy yelled. A fish, a fish, a fish. He cranked the reel even harder and then dropped the rod and pulled in the remaining line, hand over hand, until he had the squirming, hooked brook perch dangling from his hand. He was all smiles. Pure electric-pulse adrenaline pumped through his veins. I grabbed the fish and removed the treble hook. He wanted to keep it, like all kids do. I abruptly threw the fish back, and he looked at me in disbelief, like I had thrown his toy into the garbage. I bent down to him with a mean and serious and stone-faced look.

"Keep later. Now. Have fun. Catch many more," I said in my stilted Korean.

The boy snatched up the rod, strode to the water, and belted out another cast, determined to get another fix. We fished for another hour, and he caught three more smashmouth perch. It was close to dinnertime, and we walked back to the bridge.

I gave him the light-action St. Croix rod with the Shimano reel, the only spinning rod I had left, as I had snapped one two months before. I gave him a tackle pack loaded with Mepps spinners and Rapala crankbaits. I gave him a small tacklebox filled with bullet weights, lead jigs, and offset Mustad hooks, along with pouches of soft plastic worms and curly-tailed grubs. In the pack were snippers and a hemostat and a folding fillet knife. The boy thanked me many times, and I downplayed it and said, "*Gwenchana.*" It's okay. Not a problem.

I shook his hands and told him good luck and turned around. I thought of what his grandparents might think when the boy came home with hundreds of dollars of imported angling gear. They might think he had stolen it. What other explanation would there be in such a remote area? I walked back to the kid and asked him his name.

He said his name was Jae-sung. I went back to my car and found my business card. On the back I wrote in Korean: Jae-sung, Good luck fishing. James. I stuffed the card in the side pocket of the tackle pack, and we said goodbye. He waved as I drove off.

As I drove home down the Seomjin River valley, I felt happy because I had taught this random boy how to fish and given him the proper tools to do it on his own. He was poor and his family was poor. He was unlike other kids who had private tutors, piano lessons, and online computer games. He did not have those things to keep him busy. Now he had something that he could pour his young energy into with the confidence and equipment for the pursuit. I also felt happy because I had divested myself of all my spinning gear and related tackle. It was exhausting and expensive to keep track of all of the hardware. I was now exclusively fly fishing, and I enjoyed this minimalist and athletic form of fishing. To the east were the foothills of the Jiri Mountains. I had heard rumors of trout in its freestone streams deep in its interior valleys.

Two days later I drove to the bookstore. Knots. I had not taught Jae-sung to tie a lousy fishing knot. All of those lures and he did not know how to tie them on. I thought of the fish he would lose if the boy merely used a common granny knot. I bought a couple of how-to fishing books in Korean and a book of angling essays by Ahn Jung-hyo, South Korea's best novelist that writes in the English language. He fought in Vietnam and was an avid fisherman. I paid for it all with coupons that were issued for the sole purpose of supporting the Korean arts. The coupons were often given as bonuses from employers or as gifts. At home, I paged through the books and tagged passages that held information I thought was important to know with sticky notes and wrote: *Gongbu*. Study. I found a large ziplock bag and stuffed the books in there.

A day later I drove back to the bridge, and the boy was not around. I tucked the books near the boulder where he had caught his first brook perch. I knew he would be back there just like a criminal returning to the scene of a crime. The same applies to anglers. You never forget that exact spot where you caught your first fish or a big fish or a special fish, and you always have a compulsion to go back and cast a line. I tore out a page from my journal and scribbled a note and tucked it into the plastic bag.

To Jae-sung. From: Your Fishing Friend.

5 Escape Country

It was a rainy morning at dawn when the mist hung on the mountain slopes like a bedsheet over a lover and, with the slowest of movements, slid to the valley floor. In a cedar tree a magpie cackled, and down below in the rice field, egrets speared frogs that had sung the night before. In the rain, the rice fields in the foothills had taken on a color closer to neon green than their natural green. Above the treetops white flags shimmered from bamboo poles that marked the locations of chestnut trees.

This is a mountain range where pilgrims escape their secular lives to bow before a mountain spirit. In this mountain range, herbalists forage and capture its secret energy, poachers seek to reduce its wildness to dead possession, orchardists grow persimmon on its steep slopes, and the urbanites come to breathe its clean air and drink its cold water. It is here where invasions were fought with arrows and, later, partisans fought with rifles in this mountain redoubt long after the end of the Korean War.

This is a mountain range where Choe Chi-won escaped to get away from a corrupt society. Near the village of Beomwang in the Hadong district, he planted a Korean black hackberry tree that is estimated to be five hundred years old. Near the tree is a rock called the "Ear Washing Boulder," where Choe cleaned out his ears, which were tainted by the degeneracy of the outside world.

This is a mountain range where Korea's most libidinous lovers escaped to live happily ever after. In the obscene *pansori* (a musical tale) of Byun Gang-soe, the legendary cock-strong womanizer met his match in Ong-nyeo, a lewd woman with a vagina so robust she screwed men to death, and a stream of her piss was strong enough to split firewood. Byeon could handle this seductress and, with equally matched sexual superpowers, they lived in the Jiri Mountains until Byeon cut down some sacred totem poles and was cursed with a deadly illness.

This is a mountain range where the early missionaries escaped to spend the summer in hill stations. "Here, at the height of 4,200 feet, lay a number of cottages and villas, large and small, of European type spread out among the trees—these trees also leafy and shade-giving. These dwellings were dotted about all over the extensive plateau. Many European and American visitors, especially missionaries, spend their summer vacations here, thus escaping from the oppressive heat of the lowlands," wrote Sten Bergman in his 1938 book, *In Korean Wilds and Villages*.

This is a mountain range where guerrilla fighters roamed and fought and took cover. In the seaside town of Yeosu, just southwest of Jirisan, there was a rebellion in October of 1948. It was suppressed, and a thousand rebels escaped to the hills of Jirisan and joined up with local bandits. In the history books, these people are called communists, but I always speculated they were merely rural peasants enraged after centuries of exploitation. In her 1898 travelogue, *Korea and Her Neighbors*, Isabella Bird Bishop describes Korea as a land of robbers and the robbed: the *yangban* elite she calls "the licensed vampires of the country," and the rest of the population existed only to "supply the blood for the vampires to suck."

In this mountain range, on February 7, 1951, South Korean troops executed Plan Number 5. They rounded up villagers near Wangsan Peak and murdered them. Many were women, children, and the elderly. The official number of those killed is 386, but other accounts claim it was almost double that. Later that year, General Paik Sun Yup launched Operation Rat Killer, a strategy that involved making peace with the locals yet actively hunting the leftist partisans. The last of them were captured or killed in November of 1963, ten years after the 1953 armistice was signed.

This is a mountain range where the half-moon bears have escaped the slaughter. Jirisan is the last holdout of the Asiatic black bear, which bears a white crescent of fur on its chest. Only an estimated ten to twenty bears survive in South Korea, and these mountains provide enough cover to roam unmolested and safe from attacks by poachers. A team from the Ministry of Environment reintroduced two bears from Manchuria, and the goal is to develop a breeding population of fifty bears—if they are not killed off. The bears have been hunted hard for centuries to the point of near extinction. The decimation of the native population of half-moon bears is a result of *hanyak* (traditional medicine). People believe that the bear's gall bladder bile imbues one with vitality and good health.

Jirisan National Park covers 182 square miles and extends into three provinces (North and South Jeolla and South Gyeongsang), and five cities are nestled near its foothills and flatlands. It was the country's first national park, designated in 1967, and it is the largest park in the country. It is the terminus of the Baekdu-daegan mountain range—the "White Head Great Range,"which is considered the backbone of the nation: its spiritual spine. It is home to the tallest mountain on the South Korean Peninsula, at 1,915 meters. I climbed to the summit, Cheonwangbong (Sky King Peak), three times during my first few years living in Korea. There is a custom of three generations of hikers summiting the peak to watch the sunrise over the peninsula. Nowadays I do the opposite of everyone else who enters these mountains: when they hike upward, I hike downward as water flows down into the mountain valleys, and there is the home of the peninsula's south-ernmost population of cherry trout.

When driving up through the Sancheong region and approaching the national park, you cannot see Sky King Peak or any of the other major peaks from a distance. There are layers of mountains. An expansive mountain of green stretches before you, and a mountain lies upon a larger mountain, and that mountain lies upon a bigger one, and this goes on for miles.

It was the presence of half-moon bears that drew me to Jirisan during my first few years in Korea. I figured that if bears could survive in these rugged mountains, it was a place worth visiting. The southern end of the peninsula was Korea's last wild country. I hoped to encounter a bear, but my first few forays were focused on climbing the highest peaks and visiting the ancient Buddhist temples. To be able to spot a half-moon bear, I had to get farther into the backcountry and far away from the tourist hiking routes—the same protocol I would later use to discover the whereabouts of native trout.

I did eventually see two bears, but they were only semi-wild ones. I was introduced to the bears at Moonsu Temple by Ranger Kim Sung-hwe. I wandered into the ranger station in the Gurye district of the national park. At the entranceway, I looked at a poster of wild animals that live in the park. He asked if I knew anything about Korean wildlife. I told him I did, and I was interested in half-moon bears. Ranger Kim offered me a chair, and we sat down and he produced a photo album of shots taken with a motion-detection camera. Many of the bears had been photographed at night.

"They like to eat the *dol-bae*," he said. The *dol* (stone) *bae* (pear) is known as the Ussurian pear and is a small, hard fruit and inedible for humans. He asked me if I wanted to go see one.

The head monk of Moonsu Temple is Ko Bong, and the orphan bear cubs were brought to him by a friend of the temple. Their home is in a deep mountain valley, and the road to the temple is a series of steep switchbacks that leads up past the snowline. At an elevation of 3,182 feet, it is one of Korea's highest mountain temples. He once tried to release them back into the wild, but they returned to the temple. The bears slumber in cement dens in the winter, and they are let out to roam the mountainsides and climb trees. They recognize Ko Bong's voice, and he wrestles with them. They are kept in cages, as semi-wild creatures, for their own protection and for the safety of visitors.

During my visit, one bear was sleeping but the other emerged from a concrete cave and presented himself behind a wall of red steel bars. He wore a yellow ear tag with the letters M1. The bear looked very much like a medium-sized American black bear except for the boomerang-shaped patch of white fur that stretched upward from his chest to his shoulders. The steel cage gives the impression that they are a kind of freak attraction used to draw visitors to the temple, but now it is a matter of survival of the species. Ko Bong is their guardian.

In the foothills and lower half of the mountain, there is a tree canopy of chestnut, painted maples, red pine, camellia, and oaks with a jungle-like ground cover of boreal bamboo. Higher up, stone birch and yew trees grow atop steep cliffs. Yeddo spruce and Korean firs take the aesthetic shape of weather-beaten bonsai trees clinging to granite outcroppings. Off trail, royal azaleas and Korean rhododendron create thickets so dense that it is only passable by crawling through byways used by water deer.

I had fly fished the Jirisan mountain range every week for the past two years. Taking the advice of a New Zealand fishing guide, I started a guide service and would take clients fishing up here a couple of times a month. Jirisan had become my escape country—a place to get away from domesticity and civilization. The start of spring on the Korean Peninsula is signaled by the tapping of painted maples (*Acer pictum* subsp. *mono*). This usually happens every year in early March, when the mornings are frosty and the afternoons are warm. The Jirisan hill people peg spouts on the trunks of the *gorosoe* tree and make a clever network of interconnected downhill tubing that leads to a main collection tub. The sap is harvested, bottled by the gallon, and sold for around $6 a jug. It's a seasonal side hustle for the mountain folk. Koreans call it bonewater, as it cures orthopedic ailments and works as a general health tonic and cleanser.

I spent many days that spring watching a cheerful man in his midsixties tend to his sap line, which ran parallel to this creek. He waved me over, and we greeted each other in Korean. Then he tentatively spoke a few rusty sentences in English, and I replied in Korean. He poured me a cup. It tasted just like the box elder sap that I drank once while studying wilderness survival: tree sap is sterile and is an instant source of hydration when clean water is scarce. It had a clean woody taste with a hint of sweetness. He offered to give me a jug to take home, but I didn't want to lug it around while fly fishing for the rest of the day, and I didn't want to make a special downstream trip back to my car to drop it off. I politely refused his offer but gave him an opening to save face: I dumped out my water bottle and handed it to him. He grinned and topped it off with bonewater.

As the bonewater sap is harvested, there is the blossoming of the *maehwa* tree (*Prunus mume*). The tree is regarded as a symbol of endurance against adversity since its white petals bloom on bare branches set amid the melting snow of mid-March. Soon after, the cherry trees bloom, and the countryside is overrun with tourists out to view the spectacle. After the pinkish white cherry petals drop like snowflakes, the golden forsythia flowers come out, more spring rain falls, and the trout become livelier with the new hatches of insects.

I postponed a few trips up here because of the spring sandstorms. The seasonal yellow dust that blows in from China blankets the peninsula. This is what the horizon of hell looks like. An ocher dawn, an ocher dusk, and cars pass through the noonday gloom with their headlights on. Schools close, planes are grounded. Some elderly die, children are locked inside.

Denizens that must go out mask their faces like Sand People, and those indoors nurse their pinkeye and choking cough while hacking up greasy phlegm. The contaminated air is a urine-colored miasma of fine dust, carcinogenic soot, heavy metals, and dioxin. This is the cost of Chinese industrial growth. They deforested their land into man-made deserts and filled their skies with toxic smoke. The Korean people pay heavily for it. Every year it gets worse.

From the boulder that I was sitting on, I could make a long cast upstream and across to a pool of clear water where only my tippet would touch the surface. The rest of the fly line would drape itself over the smooth rocks. The water was so clear that if you tossed a hundred-won coin into the pool, you could see it at the bottom from a distance. The pool ran tightly against a small cliff with a cluster of Korean bonsai pine, making the river-left side of the creek unapproachable.

I studied the pool for a long time. The cherry trout glided through the water column, free of care until a gray-faced buzzard passed over the pool. The bird reminded me of the American red-tailed hawk: a familiar raptor seen perched along woodland edges. As the bird passed over the pool, its shadow could be seen on the streambed. The trout disappeared.

I waited and ate an apple. An old Korean man once told me: "James, in the morning is the golden apple, afternoon is the silver apple, evening is the bronze apple. You understand?"

I did. It was another version of "an apple a day keeps the doctor away." In all cultures where an apple tree can grow, they have some kind of proverb about the healthfulness of eating apples. According to the old man, the best time to eat an apple is the morning. I was later told this proverb was nonsense. I make it a point to eat an apple in the morning anyway.

Approaching on the river-right side of the stream would position an angler too close to the trout. You would be on top of them. Making a cast from the tail end of the pool, the line would get sucked by the current within seconds, and the fly would drag in an ugly and unnatural way, like a beginner water skier who won't let go of the tow rope after falling. Any mending of the line would be a crapshoot as it would be lying among the boulders and driftwood debris.

Making a cast from upstream was similar but worse, as the trout were always facing upstream and the sun would silhouette your form. The best

Fig. 7. The trout creeks of the Jiri Mountains are a maze of tumbled boulders upon a steep gradient. Photo by James Card.

option was to back off and cast from a distance, and the only way to do it was to climb atop a big boulder that sat at a forty-five-degree angle downstream and off to the side of the pool.

The morning mist still hung in the valley, but the rising sun was quickly burning it off. It would be a hot day. I used a four-weight rod and a long leader that narrowed down to an 8x tippet. I would have preferred to use a lighter rod, but this creek was so varied and complex that I felt I needed a stronger rod to handle these situations—in this case to make a fifty-foot cast into a slight upstream breeze and time it so only the leader and tippet would touch the water.

I took a brown elk-hair caddisfly that I had blackened with a permanent marker and tied it to the tippet with a Harvey knot. Albolene is the same ointment that boxers use as a lotion to cut weight before a fight, and it is the best and cheapest floatant around. One jar will last a lifetime, and I dabbed a speck onto the caddisfly. The trout were feeding but in an opportunistic and casual way. I had observed only two rises since I had approached the pool, and they were the common black caddis. The other trout, which were not rising to the surface, dipped and lunged toward the water pouring into the pool, presumably picking off nymphs getting washed downstream.

The uphill side of the boulder had a rock that I used as a step to scale and shimmy my way to the top. I left my rod propped against the boulder and picked it up by the tip and brought it up hand over hand. It was like standing on the roof of someone's garden shed. I had to make the throw with minimal false casting. I did not want to risk any extra movement that might spook the trout. Stripping out lots of line, I let the fly line dangle against the boulder in loose loops for a quick pickup.

It would be a matter of one false cast to get most of the line airborne and into a backcast, and then a quick, hard doublehaul to shoot the remainder of the line forward with the leader and tippet only touching the water. I edged forward on the boulder, right up on the curvature with my boots gripping the granite.

I made the cast, and the caddisfly landed about where I wanted it. It hung in an eddy near the bubble line, swirled, and righted itself. I watched a trout elevate itself through the water column, and it took the fly in a slashing movement. I counted: One. It was to make time for the turn and then set the hook. The fish was on solid, and that is when I fell.

It was the first of three hairline-fractured ribs that I would have in my life: one after falling from the boulder, one from a bar fight in Changwon, and another while taking out the garbage and slipping on a very mundane patch of ice.

I fell face-first into a garden of boulders the size of bowling balls. As I felt gravity's uncaring pull, I leaned back and skidded down the boulder until there was nothing left to ride and I flipped forward. I remember letting go of the rod: *do not land on the rod.*

It knocked the wind out of me, and I rose on my knees gulping air and panting like a lung-shot beast. Breathing and breathing, and the air came back after a minute of panic. I felt my face and looked at my hands. No blood. I touched my body: elbows, knees, neck, dick, shoulders. I gently touched my skull for bumps. Everything fine. Just my torso felt torqued and hammered. I kept breathing, but my breath was on the shallow side.

The world was a brighter place. I looked up at the sky. A perfect day, high clouds, sunshine. I looked around. A water ouzel landed on a rock in front of me. It is a dark-colored perching bird with waterproofed plumage and nictitating eyes that allow it to walk underwater and hunt aquatic insects. I thought of the birds as a companion spirit as they zipped from pool to pool. My polarized glasses had settled among some rocks. I crawled over and fetched them. The ouzel took off. I put my glasses back on, and then I heard the clicking.

I stumbled over to the rod. The line was now tight, and the tip of the rod dipped and ticked against a rock. The trout was still on. I crawled over and pinched the fly line against the rod and felt the movement of the trout like a blind seer laying hands upon a telegraph line and understanding what is being communicated.

I pointed the rod toward the trout and lifted it up, and I felt the fish but not the fight. I reeled in the line as I walked to the pool. The angle of the fly line didn't make sense until I was near the water's edge. The line was wrapped around a cedar branch that flexed like a bow in the water, and the trout was thrashing in the whitewater as if it were hooked on a trot line. I untangled the loop, reeled in the slack, and brought the trout to net. It was a twelve-incher, a very nice trout for a stream of this size. I unhooked the fish and released it into an eddy. It did not dart down to the depths like it should have. It finned in the slow water, lethargic. I watched it for a couple of minutes. It would not live. It had exhausted itself when hung up on the tree branch in the rushing water. I snagged the trout and pressed it against a small boulder. Then I took my Canadian belt knife and sliced a notch across its spine, right behind the gills, severing its spinal cord. I soaked a handkerchief in the water and wrapped the trout and tucked it into my vest.

I always released fish on this stream, and this was the second trout I'd ever kept from these waters. The first was when a client was posing for a grip-and-grin photo and dropped the trout into a crevice. It was deep and both of us could barely grasp the fish. We were up to our armpits reaching into the hole. I finally used my hemostats to gain a few more inches of reach and plucked out the trout by the tip of its dorsal fin. It was dead from flopping around in the gravel-bottomed hole. The client was staying at a hotel and had no way to cook it, so I brought it home and ate it.

My ribcage hurt but I was already thinking about how I would cook the trout later that evening, but then I decided it would be better for breakfast. It was a rare event to keep a cherry trout. I had caught and eaten many freshwater fish but not the cherry trout from the Jirisan streams. The habitat and population and balance were too precarious. This had to be a special meal. I decided on Eggs Shannon: fry the trout, top it with a poached egg on creamed spinach and slathered with Hollandaise sauce.

By now the mist was gone in the valley. In my field notes I always wrote "Geolim Valley" as "Gollum Valley," after the crawly scoundrel from *The Lord of the Rings*. It was a place that the traitorous little bastard would have liked: boulders, crevices, and a stream for snatching raw fish to eat. The creek that ran through the valley, the Naedae, I also spelled in my own way: as "Nayday." It rhymed with *mayday* and just had a nice ring to it.

My breathing was raspy when I took a breath, and my ribcage had a dull ache. I continued fishing but it was slow going. If you were in top condition, boulder jumping Nayday Creek was a form of parkour with a fly rod. There were gaps to leap from boulder to boulder, and they were always wide enough to make you uncomfortable—enough to make you pause and doubt yourself until you size it up, steel your nerve, and make the leap. Every step involved rock hopping, with so many rocks that wearing waders wasn't even necessary—you could tiptoe and scramble atop the boulders without getting a foot wet.

I have done both, with waders and without. The only reason waders were necessary was for making shortcuts when the obvious path heading upstream petered out. Then you would have to make a long detour, like hiking an extra hundred-yard loop to get around an obstruction to progress upstream ten yards. Some clients wanted to wear waders for comfort, and I always provided them. Or sometimes it was better to wear waders as rain gear because the weather was cold and miserable.

The best combination is doing a mix of rock hopping and wet wading, and the best clothing combination is polypropylene long johns under a pair of quick-dry cargo shorts. I quit wearing wading boots a long time ago, as I've never found a pair of boots that can withstand the underwater abuse of gravel and silt and sand and mud constantly immersed in the seams and fabric. The never-ending soaking and drying cycle always leads to breakdown, and the sole is usually the first thing to delaminate from the boot. A few times I have walked back to the car with a floppy boot sole.

The best wading boots are a pair of hiking boots that can be picked up at a thrift store or a rummage sale for a few bucks. Use them until they start to fall apart, and then throw them away. Always have two sizes: one for wearing with regular socks, for wet wading, and another pair that is about two sizes too big, to wear with thick neoprene stocking-foot waders.

I've found the felt-sole wading boots and ones studded with grips to be mostly unnecessary. There were only two places in my entire life where the rocks were so slippery that I became terrified. One was the Namdae stream in the Gangwon Province, and the other was a small creek on the South Island of New Zealand. Overall, both rivers were not very slippery, but each had a small section where the rocks were covered with algae slime, and I happened to wade into it. I fell both times and was lucky I did not cock my head on a rock. I saved myself by doing a contortionist dance of spastic moves. I think of the outdoor writer and former diplomat Datus Proper, who drowned while fishing Hyalite Creek outside of Bozeman. It is thought that he slipped on a rock and hit his head. It is true that a person can drown in very shallow water.

The Nayday is one of the most complicated and weirdest trout streams on Earth. The valley is a tumbled mess of boulders, as if God hurled a handful of polished rocks into a steep canyon with a creek in it. It does not flow on a gently descending and meandering horizontal plane like most small streams. On a normal stream, you simply gaze ahead to survey the upcoming riffles and pools. On the Nayday, the next pool is above your head and ten feet away behind a screen of jumbled boulders. Rarely can the water upstream be seen. Most rivers and streams seem horizontal to the human eye, but they are not. After all, all water flows downhill so there is an eventual gradient, and some are more noticeable than others. So anglers look upon a river with a linear perspective: the river is a two-dimensional flat line, and there is upstream and downstream.

But the Nayday's sinuosity is like no other: it is three dimensional and more vertical than horizontal. If two school kids were to replicate stream flow using a garden hose, one kid would dump the hose on a sandy patch of land and point out how the velocity of water erodes the sand and makes a channel. The other kid, tasked with replicating the stream flow of the Nayday, would dump the hose on top of a smashed rock-hewn stairway and say: it is more like this than that. One moment you're casting into a clear pool, and the next you are scaling a boulder the size of a dump truck, sliding down a crevice, and then making a small leap into some pocket water, wading through, and backing up into a position to survey where the next cast will go. Some flies are presented as if you were casting from the top of a two-story house into the swimming pool below. It's a thinking man's creek. With the steep gradient, you have to scramble over a boulder or scale a rock face, and then you look at each run and pool from a different angle. It forces you to slow down and think. Each plunge pool is a case study.

The Nayday flows through this stochastic maze of tumbled boulders, but a simple pattern emerges. Most trout streams have a riffle and pool sequence. The Nayday has a waterfall and plunge pool pattern. The water is more violent because of the vertical-drop velocity and more beautiful because everyone loves a waterfall, whether it is Niagara or a three-footer. There are other water forms besides the plunge pools and waterfalls. In one spot the creek disappears entirely, as if it went down a drain, but navigate around a huge patch of boulders and you find a pool twenty feet below. Some of the rock surfaces are marked by perfectly round holes that look like the bore of a cannon, and they were formed by centuries of swirling, scouring gravel.

Between these chaotic boulders are chutes, cataracts, pools, glides, and riffles, and that is where I discovered the cherry trout in this clear-water geological jungle gym. The headwaters come out of the Jiri Mountain Range, and the local government had created the country's southernmost trout stream by stocking the Nayday with *sancheoneo* a few years earlier. The Korean name translates to "mountain stream fish." Native to the Korean Peninsula, Japan, Taiwan, and the Russian Far East, *Oncorhynchus masou* is called either the cherry trout or the masou trout in the English language. The Japanese call it *yamame*, which means "mountain lady fish."

Sometimes it takes the patience of a Buddhist monk seeking enlightenment to spot a trout even though the water is so clear. They are small, usually less than twelve inches, and their colors are an olive green dappled with black specks over its lateral line. The fish has a greenish gold back becoming yellow on the sides, and it keeps its parr marks throughout its lifespan.

The creek is a paradise for technical presentations. I use steeple casts and Galway casts just as much as conventional forward casts. In one spot I made a bow-and-arrow cast in a gap between two boulders, much like an archer shooting through an arrow slit in a medieval fortress. Other times I leaned over a boulder using it as cover and performed a tenkara-like

Fig. 8. Landlocked cherry trout, *Oncorhynchus masou,* are the jewels of the Korean mountain streams—not just in beauty but also in rarity. Photo by James Card.

dapping of a dry fly. One trick was to cast the fly line so it would land on top of boulders with only the leader and tippet in the water. As the fly was about to drag, I flipped the line off one rock and onto another to get a few more seconds of drift time. I called it the "boulder mend."

Sometimes I brought kneepads because there was no other way to get closer to the creek than to crawl across the scree and, from there, make a low-profile sidearm cast, where the rod is cast parallel to the water's surface. There is no one to blame but yourself if you bust a pool, and the consequences of your errors are readily visible and invariably instructive: the trout you were once watching are gone. You must eat your failed cast and move up to the next pool.

The tail end of a pool is often overlooked. The pole position of the biggest trout in a pool is where it has the best protective cover while also getting the most food. This is usually at the head of the pool, and that trout has first dibs on all the food getting flushed into the pool. If that trout is caught and killed, the next biggest trout will take that spot and become the pool's alpha. However, there is sometimes a trout that is an outlier and finds a niche at the very end of the pool because the current funnels the leftovers and missed food and the trout at the tail gets a concentrated food supply. The trout in this position is never the biggest trout in the pool, but it is always one of the larger ones.

The line getting sucked down the waterfalls was one of my biggest annoyances. Standing at the back end of the pool where the water plunges down into the next waterfall, you think you are being sneaky and smart. You think that if the trout are facing upstream, they can't see you and you can work your way up the pool-run presenting your fly to the local residents. But the drag on the line sucks harder than you think, and when you leave your line on the water a few seconds longer, as you hop to another boulder, you see the miserable sight of the fly line getting vacuumed into a waterfall of mangled boulders. They are eaters of flies. I recovered some by reaching up deep into the frothing water to unsnag a fly caught on a bit of debris wedged between the boulders.

Stealth and strategy are of the utmost importance. Your rod cannot be flashing back and forth high in the cone of the trout's vision. Rocks do not sway in the breeze. In this clear water that runs like liquid diamond, they are hypersensitive to movement, more so than their cousins that live in forested streams with canopies overhead. In the streams of the north near DMZ country, I can wade and cast and casually work my way upstream and have a very pleasant day. Not here. I have to climb up and over another mass of boulders, and a new plunge pool is revealed. I spot more trout in the water column, and a new chess game begins.

The Jiri Mountain range is composed of granite gneiss—speckled boulders of gray and tan—and I always wore a gray and brown checked shepherd's shirt that blended well against the boulders. My Filson fly vest was gray, and that worked well enough. Without overhead canopy cover, the trout use the sun and shadows as cover. At noon the sun directly overhead makes them very nervous. As the sun moves across the sky, the trout move with the shadows. There is a rule: the darker the water, the better.

I caught five more cherry trout in the next few pools and runs, but I decided to cut the trip short. My ribcage was bruised, and rock hopping only made it worse. My ballast was gone. The boulder prairie I often fished at the end of the day was a take-out spot—the last chance to catch one more as the sun set over the western peaks. It was the flattest stretch on the entire stream: a huge, flat ledge covered with flowing water and studded with fast chutes, mirror-like placid spots where current runs swift but smooth. Sometimes the water is golden or pewter or silver or like a mirror. It was a time of day that photographers call the golden hour—when the brightness of the sky matches the brightness of the diffused light on the water. It is mirror-like because of the position of the sun at the end of the day. At this time of day, the water is always blinding. High-quality polarized sunglasses will give an angler superhuman vision: the ability to see deeply into the water. It's even better when the water is clear and the sun is out. At the close of the day, as light diminished, the sunglasses became less and less useful, and it was time to let go of the x-ray vision and just perform a squinting analysis of the water ahead. I often found at this time it was best to crouch low and look at the water with my gaze parallel to the surface. Nearing dusk, this was a time when the trout were active and rising, and it was better to watch for the rises rather than the trout themselves.

I picked up a goat trail that led to the mountain road where guesthouses and tile-roof farmhouses were dug into the steep slopes. I passed by an angelica tree—also known as the devil's walking stick—and sliced off the *dureup*, the fragrant shoots that emerge in the spring. They would go with the trout, and they are cooked like American fiddleheads. There was a beginning and an ending ritual. Upon parking my car, I loaded plastic bottles of OB Lager into a basket net that I had repurposed as a beer soaker. I hiked down to the creek and added a rock to the basket, tied it off with paracord, and lowered it into the pool. I would discreetly knot off the running line to a tree branch. At the end of the day there was always cold beer to be drunk, and it was always a pleasant ritual to crack one open with my clients and talk about the highlights of the trip.

I doubled back down to the creek and found my net basket of beer unmolested. Back at the car, I performed my copacetic ritual of sipping on a coldie, stowing my gear, and writing in my field journal. I put the trout in a small cooler and submerged it in cold creek water to preserve it for the ride home. The beer made my ribcage feel better, and I decided I wanted a few more. These mountain trout streams were important to me, more than I could have imagined. They were my place to escape from the mundane madness and frenetic pace of modern South Korea. The trout streams of the north near the DMZ were such a long drive that I needed a few days at a time to make it worth my while to make the trip. I often stacked clients' trips back-to-back up there and then set aside another day or two for me to fish alone and explore new water. But the Jirisan streams were nearby, and I found myself heading up to the mountains two or three or four times a week. Soon, these waters would no longer be a place to escape to as the urban hordes would descend into the valleys. As I drove home, I considered my fishing options during South Korea's summer tourist season.

6   Piscine Doppelgängers

On my first trip on Korean Air, there was a flight attendant who looked like Stacy, a beautiful woman I knew in college. She later moved to Long Island and became a dentist and married a dentist and had kids with perfect teeth. The flight attendant was a dark-haired Korean woman. Stacy was a sandy-blond Caucasian. They were identical twins but that was impossible. The flight attendant was a Korean version of Stacy, and Stacy was a blond American version of the Korean flight attendant. The shape of her face, the angle of her cheekbones, her smiles, her lithe arms, her delicate touch, her pert breasts, her skinny legs, and her tight ass all matched up. The resemblance was uncanny, and I could not take my eyes off her.

The husband of Mrs. Kim, my first Korean co-teacher, looked like Quanah Parker. When I showed him a photo of the Comanche leader, he grew silent and studied it for a long time.

His wife walked over, took one look at it, and gasped. "Oh my God. That's you!" she said. He asked if he could keep the picture. He grabbed a pencil and asked me how to spell Quanah.

While working at the Daewoo shipyard, I had an older student nearing retirement who was a Korean version of Abe Vigoda, who played Salvatore Tessio in *The Godfather*. I had another student in a high school class who appeared to be a young Mike Tyson. Despite the difference in skin color, their facial features, smile, and body structure were identical.

After a while I concluded that the concept of doppelgängers transcends races, species, cultures, and continents. The origin of the word is German, and it means "double walkers." It is the idea that somewhere out in the world people have look-alikes who are biologically unrelated to them. Doppelgängers have been recorded in myth and literature for centuries.

I also started to notice doppelgängers in the natural world. The pine trees of northern Asia look like the pine trees that grow in North America. This isn't too remarkable. But then you notice this with other trees, too, like oak, larch, and rhododendron. Mostly the same but a little bit different. It is similar with birds, insects, and mammals.

Sometimes the similarities are more behavioral and coincidental. Take for instance the American wood duck (*Aix sponsa*) and the mandarin duck (*Aix galericulata*) found in East Asia. They are the only species that make up the *Aix* genus. Both are considered to be the most beautiful of all waterfowl on their respective continents. Both nest in woodlands, and their feeding and breeding habits are very similar. Side by side, their colors and plumage are different, but their commonality is identical: they are beautiful sights to behold that are found in pretty water that flows through the forest.

What really caught my attention were two species of freshwater fish in Korea that happen to resemble two of the greatest saltwater game fish in the world: the bonefish (*Albula vulpes*) and the Atlantic tarpon (*Megalops atlanticus*). These two species gave birth to the sport of saltwater fly fishing, and entire travel industries have been built around them. There are bonefish lodges in the Bahamas, and legendary fishing guides have dedicated their lives to chasing tarpon in the Florida Keys.

These two species of saltwater fish, which represent the ultimate in fly-fishing adventure, have freshwater doppelgängers living on the Korean Peninsula. They transcend not only continents (Americas versus Asia) but also water (fresh versus salt) and climate (tropical versus temperate). If natural history plates of these two species were put side by side, the Korean freshwater species would look like lost cousins of the saltwater fish. The body shape, bone structure, and fin positions are nearly identical. They are just smaller freshwater versions.

I heard of Korean anglers catching these fish with lures and live bait, but if these species were studied with fly fishing in mind, they might be the next great undiscovered game fish to take on a fly rod. These were the Korean Peninsula's known unknowns—fish that were known to exist but have no known fly-fishing techniques to catch them.

Mighty Half-Beards

At a predawn hour I stopped in neon-lit downtown Changwon to get my breakfast at a carny-like food stall tent called a *pojangmacha*. They are the only places open at 4:30 in the morning for a soju sunrise. The bleary-eyed owner welcomed me in. The only people in the tent were a woman wearing garish makeup and thigh-high leather boots and a man wearing a work shirt worn by the autoworkers at the local GM Daewoo factory. According to newspaper reports, he probably just got laid off. They shared a plate of grilled eel.

I ordered some *soondae*, a Korean soul food of blood sausage, liver, and chitterlings. The man noticed my fishing vest and said, "*Nakshi, nakshi*?" The Korean word for fishing. I told him yes, I'm going fishing.

"Sea fishing?" he asked in English.

"*Anio. Nunchi nakshi*," I replied. No, I'm going fishing for barbels.

"*Ah, nunchi*." He extended his forearm, flexed it hard, and pinched the inside of his bicep crease indicating how long they get. "Strong fish. Where are you going fishing?"

"The Nam River." And so began the twenty questions. Around the eleventh he invited me to sit down and trade shots of soju. He was loaded and really wanted to be my drinking buddy. When my order was ready, I paid for it and got out of there fast. "Catch a lot," he hollered after me.

Back at my car, both windshield wipers were festooned with sex cards for massage outcalls and late-night companionship. They featured beautiful nude women and guaranteed a good time. I plucked them off the windshield to give to a Canadian friend who collected them. His collection numbered in the hundreds, and he kept them in baseball card binders. He mailed extras to friends around the world.

I drank coffee and ate on the way. To make time and beat the speed cameras, I rolled down the window and blasted the lens of the camera with a one million candlepower spotlight. If you are speeding, the camera takes a photo of you in the vehicle, and it is mailed to your home with the fine you must pay. The passenger side of the image is blacked out. Legend has it, a politician was driving around with his mistress, he got busted speeding, and his wife was the one who opened the ticket when it came in the mail. He put an end to the passenger side being visible in the picture.

Forty-five minutes later the sun bled over a craggy bluff crowned with Japanese red pines and choked with kudzu. I idled the Daewoo LeMans atop an embankment road and looked over a section of the Nam River. It looked beardish. That's what I had come to call the water I was looking for: beard water. The kind of water where half-beards could be sight fished for. I was outside the city of Jinju, and on the river was an ancient fortress built in the Koryo Dynasty. It was here upon a rock where Nongae, a *kiaseng* (similar to a geisha), seduced a Japanese general, embraced him, and threw herself into the river to drown the general. It is believed that the rock changes position after a long period of time, sometimes pointing shorewise and at other times riverwise. When the rock is leaning toward the shore, it is believed to be a time when war could break out. I always saw the rock leaning toward shore.

The embankment roads were a dangerous maze as they did not usually show up on maps. They were skinny cement roads with the riprapped river on one side and a dirt bank on the other that dropped down to rice paddies or other ag fields. Some were red-dirt roads, *hwangto*, red earth, as Koreans called the red-yellow loess soil thought to have medicinal qualities. Mostly they were used by the farmers with their two-wheeled rice paddy tractors. They were interlinked but sometimes they would dead-end. These one-tracks were barely wide enough for the passage of a medium-sized vehicle.

There was usually no way to turn around, and many times I had to drive backward with perfect precision or one of my wheels would peel off the slab, the tire would drop, and the rear axle would crash into the edge of the cement. Sometimes I could barrel out of this, other times it was too steep. I started carrying three bottle jacks to get myself out of such a fix. Once the car was jacked up, I wedged a rock under the suspended wheel, lowered the tire onto the rock, and drove back onto pavement. The other bitch of using these roads to scout new water was the farmers going to or returning from their fields with their rice tractors. The way I looked at it, I was the interloper and this was their access road to their fields. I always had to be sure not to block the road, and sometimes I had to ditch the car a mile away from the water I wanted to fish to keep the road clear.

I was trafficking in hunches by studying the maps and watching how the river flowed. It was the exact opposite of what my angler's mind had been trained to do, that is, to look for deep water, drop-offs, strong currents, and runs. Instead, I looked for the shallow slack-water flats. I am a water-watcher looking for that ephemeral patch of liquid that I know holds the fish. I can always recognize it when I see it. I took a last drink from my duct-taped coffee mug and got out of the car and looked over the water with my Leupold pocket binoculars. By now the sun was shining on the Nam River, and it made for better sighting. After a couple of minutes, I spotted what looked like a half-beard cruiser, and if it wasn't a half-beard, it was a pretty damn big fish. I grabbed my rod and vest and hiked down the riprap to the river's edge.

I did not have to wait long. The barbels ghosted in, they ghosted out. They are gray shadows in the shallows. The first pair glided in and more singles and pairs. The next one came in at an odd angle, almost directly behind me, and I froze. The only thing moving was my rod tip because my hands were shaking from the excitement. This was Gestalt therapy with a fly rod.

One swims three feet behind me, and another stops for a second before me and fades. The half-beards are in the shallows for a reason. The shallows provide no protective cover whatsoever—they are there to gorge. In the shallows they are exposed to predators, and they are a nervous fish, glid-ing here and there to forage but also making themselves moving targets. The sound of footsteps on river stones is the worst and will spook nearby barbels. You must control your shadow and think about where it will cast before making the next step. Wind is the enemy because the surface riffles obscure your ability to sight the fish, but the riffles can also be used as cover to make a cast or move positions.

A large fish swims off twenty feet to my right. I sidearm cast and lay the fly on the edge of its peripheral vision and let it sink and then give it a twitch. The barbel keys in on the movement, investigates, eats. The fly line zaps out of the water, and the fish stabs upstream. I lean into the old six-weight St. Croix and put some bend into it.

I capture the horse-faced fish with a net I use for cherry trout. It is way too small for the job; half of it sticks out, an uncouth slab of flopping fish muscle. Its body is as thick as the fat end of a softball bat, and I pop off the hook and let him blast back into the river.

Looking up at the nearby highway bridge, I realize I have an audience. A morning person on a three-speed bike leans against the guardrail, and he waves and pumps his fist in the air and then holds his hands apart to communicate: big fish.

The Amur barbel is distributed throughout the Korean Peninsula, Japan, China, and Siberia. The other species of the *Barbus* genus are scattered throughout Europe, Mesopotamia, and central Asia. The word *barbus* is Latin for "beard," referring to the whisker-like appendages near the fish's mouth, and thus, the Amur barbel found in Korea, *Hemibarbus labeo*, means "half-beard"—its barbs don't get that big.

In England, some anglers nicknamed the subspecies of barbel that live in their waters "kings of the river." I called the half-beards I pursued in Korea the poor-man's bonefish. This is because they are almost identical in appearance to the famed bonefish of the tropical saltwater flats—their shape, size, fin structure, and appearance all match up.

At first, I could not catch them. I threw everything in my fly box at them. I read everything I could about them in Korean fisheries literature, which was not much. But if they looked like bonefish, why not fish for them like bonefish? I had never caught a bonefish before.

I read everything I could about fishing for bonefish. My self-invented course of study was absurd: learn how to catch a freshwater fish that lives in another part of the world by learning how to catch a saltwater fish that lives in an entirely different ecosystem and climate. It would be like learning how to fish for muskies in Minnesota by reading about barracuda fishing in Thailand.

But there was method to the madness. Besides appearance, they have more in common than I thought. Bonefish are caught in shallow saltwater flats. Barbels are sightfished on shallow river flats. Both are bottom-feeding fish with downturned mouths for sucking up creepy crawlies on the river bottom or the salt flats. The angling experience is similar—the fight is muscular and violent and they are sprinters.

Developing a working knowledge of this fish took all summer. I would watch the weather. The day had to be calm and no wind. Korea's nickname is the "land of the morning calm." That is what I was looking for. The wind would riffle the surface water, transforming the windowpane surface into a television signal that never comes into focus—just endless static and blur. It is possible to fish for them, but you are fishing blind and the fun is gone.

Fig. 9. The half-beard barbel, *Hemibarbus labeo*, are targeted using tactics identical to fly-fishing for bonefish. Photo by James Card.

There is no athletic pursuit. You might as well sit on the bank, wet a line, and crack a beer if you are going to do that.

I called the fish different names at different times: barbels, gray ghosts, half-beards, K-bones, and scarfers. The half-beards are scarfers, not strikers. They don't have that magic take of a trout, nor do they powerfully inhale their prey like a largemouth bass. They scarf up their prey—crustaceans, worms, nymphs, and even minnows—like wild boars rooting for truffles.

There are no specific flies to use for barbels other than what I called creepy crawlies or hairy-and-scary. The bigger the better. Anything less than size 12 is too small but could still work. You want the fly to be seen from a distance because of the offset presentation. Damselfly, dragonfly, stonefly, and mayfly nymphs all worked well. I've probably caught most barbels on black woolly buggers, so anything that had leech-like undulations worked well too.

I've told many clients that if I were tasked with curating a global survival kit—the kind of kit that would help you survive in the Peruvian jungle, the Mongolian steppes, or the Canadian Rockies—the kit would have bead-head woollybuggers as the first choice of fly or lure. But I also point out, you might as well throw in a packet of dry flies, wet flies, and nymphs since they are nearly weightless.

On the Nam River, there was a tree that was a roost for Chinese egrets (*Egretta eulophotes*), and it was near a river flat that was good hunting grounds for both the egrets and the half-beards. I noticed the barbels would cruise and congregate under the tree, and I learned that they were eating the bird droppings. They waited for the squirt and splash. I tied on a white woollybugger and always caught fish there. They were like the black pigs of Jeju Island, where the swine pens are positioned under outhouses and the hogs lap up freshly squeezed human turds. Among connoisseurs of Korean grilled meats, this is the finest of all the pork in the country.

The mouths of half-beards are rubbery compared to the rough cartilage of other game fish, and hook removal was difficult. It was like a fishhook stabbed into a racquetball. Once I noticed this, I pinched down the barbs on all of the flies in my barbel box, and pulling out the hook was much easier and left the fish in better condition upon release.

The one key that all good barbel flies should have in common is weight, but not too much weight. The weight can come in the form of a bead-head or tungsten wire wrapped onto the shank of the hook, but too much weight risks spooking the fish. Sometimes, depending on the material from which the fly is made, just its being saturated with water is enough to make it drop fast enough.

If you see a cruiser and make a throw where the fly lands five feet away, you do not want the fly to resemble a pebble tossed into the water. It is too much and the fish might flare. However, you do want enough weight to get the fly quickly to the bottom. If the fly is too light, it will hang in the water column and slowly sink and the barbel will pass right under it or next to it. It is a bottom-feeding fish, and the fly must be presented from the bottom up.

There is a fine line between a pebble hitting the water and a slight "bloop" like an insect falling into the river. The latter is what is desired, as occasionally I've seen the barbel come to investigate the tremor. Usually, the small pebble-like splash of a heavy bead-head will set them off to the other side of the river flats. When a barbel does take the fly, it is pig-like and unthinking. It scarfs it up. Often the fish is hooked well in its rubbery lips, and the hook will hold during its initial run and during the ensuing rodeo.

Everything is done delicately and slowly except for the casts: one false cast to get the line in motion, a double haul, a follow-through, the placement, and done. Once the fly sinks, it is animated as it is meant to be: painfully slowly. Keep the rod tip pointed directly at the fly, with no slack and a direct tight-line connection. You need to tease the fly with small strips; add a shimmy, a gentle shake, a puff of sediment digging the fly into the sand or muck, a light twitter, and make that marabou pulse and undulate.

Another thing to factor into the presentation is the sink rate of your fly. Give it a count: one Mississippi, two Mississippi, and figure out how long it takes for the fly to reach the bottom in different depths of the water column. It is a quick mental calculation that must be made upon a half-beard sighting: the direction the barbel is traveling, the cast to be made, and how much to lead the fish and the fly entering the water without spooking the fish and sinking to a spot where it will be noticed by the cruiser.

I've caught barbels that were almost two feet long, and a six-weight rod works best—for fighting and also throwing weighted flies for distance. The presentation is the key, and where the fly lands in the proximity of the barbel means a look and a hookup or a rejection. The fly must land off to the side of the fish that is closest to you. If it lands on the other side of the half-beard, it means that you will retrieve the fly toward you and toward the fish, which is an unnatural act as no prey swims into the mouth of a hungry predator. The trick is to mimic the fear and flight of a small creature that intuitively knows it is about to be eaten.

Sometimes the take is aggressive and sometimes it is subtle, and then there is one in the middle, where it feels like your fly was sucked up by a low-power vacuum cleaner—nothing at first and then a slow, steady tension and tightening of the line. Either way, slack must be out of the line, and as soon as it appears the half-beard is mouthing the fly, it is time to strip-strike. Just because the barbel seems to have the fly in its mouth doesn't mean it is solidly hooked.

When setting the hook, keep the rod pointed at the fish, give the line a hard and fast pull straight back, and drive the hook into the rubbery mouth of the half-beard. Lifting the rod and yanking it back over the shoulder, like a pro bass angler reefing on a largemouth, doesn't work. It is too slow and sloppy, and if you miss the set, your fly goes airborne. With the strip-strike, a miss results in the fly skidding forward a foot, and the barbel may pick it up again.

It's best to fight from the hip and keep the rod angled to the surface. There isn't much of a point to raising the rod much higher other than lifting the line to clear snags and rocks where it might get tangled.

The drag is set lightly but enough to keep tension on the fish. Set it and forget it. You never want to dick with drag adjustments while a powerful fish is on the line. Your focus needs to be on reading the fish. A light drag can be handled by feathering the spool; a drag that is too tight results in a broken-off fish.

The thing to remember is that this is a river and not an expansive saltwater flat in the Bahamas. The river has deep holes, snags, strong currents, and a tumble of rock and debris on the riverbed. Chasing the fish down is often not possible. The chance of stepping into a deep hole is too great along with a sandbar collapsing and getting sucked into the main channel.

Sometimes the half-beard will run back to me, which creates a belly of slack line. The first instinct is to raise the rod to the sky to get the line tight again. This never seems to work. It's best to strip in line as fast as you can. Don't even bother with the reel or think about going back to fighting the fish from the reel. The only concern is to maintain a direct line connection with the fish. Once the line is tight against the fish, you can keep tension on the line by pinching the line down against the cork handle while reeling all the loose line that is now piled up by your feet. That line needs to be cleared. If the barbel were to bolt again and you were accidently stepping on the line, the barbel would be like a chained-up spastic dog: the dog charges toward some excitement and, at the end of the chain, is violently jerked back. With the barbel, the chain is the tippet and the barbel will break the tippet.

Most of my casts to half-beards were between twenty and sixty feet. The long casts were very rare, and the short casts were nerve-racking—the fish were simply too close and they would spook at the plop of the fly. The long casts were possible only when the visibility conditions were right: you cannot spot a feeding or cruising barbel from that far away unless you are elevated. Sometimes the sunlight, the calmness of the surface, and the angle and depth of the river would all come together and a barbel could be spotted from a distance, but it wasn't often. Polarized glasses are essential, and the brim of my Doosan Bears baseball cap helped cut the glare.

To cast from an elevated position, I mulled over inventing a kind of stilts, like those worn by circus performers. The stilts would have special clawed feet to grip the rocky and slippery riverbed. They would be strapped to my ankles, calves, and thighs, and I could wade over the flats like a lumbering giant. I remembered as a boy I built some stilts, as a Cub Scout project, out of scrap wood and got pretty good on them, even being able to do a jerky sprint. But it would still be very unstable. I considered supplementing the stilts with a huge wading staff and, when casting, I would strap the staff to my chest and I could lean into it for greater stability like a human tripod. I tabled this idea.

I experimented with a wooden stepladder. It had only three steps, the kind you would use to get something off a high shelf. It worked excellently and exactly how I predicted it would, and I could sight half-beards from a great distance—even farther than I could cast. I grew to hate it. It was clunky and heavy, and it was annoying to haul to the river and store in my car. It got heavier toward the end of the day because the wood absorbed water and the bottom step would become waterlogged. It was also noisy to set up, and I despised anything that took away from my stealth. Sometimes the loose fly line would become pinched in the hinge of the ladder and need to be cleared. I also almost fell off the ladder when I hooked into a half-beard. It became a personal rule to back down the steps and get sure footing on the river bottom when a fish was hooked—and that was one more thing to focus on during the first adrenaline-rush seconds of the fight.

The trade-off with the ladder was stealth. It was clunky to move around, and stealth was everything in the half-beard game. I tried to keep my shadow behind me. Watch the sun, watch the shadows. Midday creates short shadows, but the best times to fish—sunrise and sunset—cast long shadows. My favorite footwear were worn-out leather moccasin-style chukka boots as they let me get a good feel for solid footing without much disturbance. Eventually they disintegrated and I replaced them with an old pair of trail-running sneakers, which worked well too. I often waded with hard-soled work boots, but on the half-beard flats they were too clunky for quiet stalking. Because sound travels better through water than air, one cluck of an underwater rock and the barbes will scatter into deeper holes. Everyone stumbles while wading, and stumbles create waves, and waves scare skittish fish. It was much like still hunting for whitetail deer: look down to where your next footstep will be and check for crunchy leaves or twigs that will snap. Adjust and place the foot by slowly rolling it onto the ground. It can be heel-to-toe or toe-to-heel. It depends on the terrain. Just a slow and precise distribution of body weight to the ground. Stop. Look. Listen. Wait. Take another slow step. Repeat for one hundred yards.

While silent-stalking the river flats, I found lotus candles washed up into nooks along the bank. They were released during a spring celebration for Buddha's birthday, which falls in April or May depending on the lunar calendar. There were hundreds of them, and I filled my backpack to reduce the litter in the river and brought them home to burn in my courtyard for evening chill-out sessions while sipping on beer and slow-smoking meat on the grill.

I developed two tactics with the ladder. One was to set it up immediately in a section of the shallows that offered the best casting and sighting position from every angle and then leave it. I would explore the edges of the river flat and then come back to it and fish from the ladder like a perched gargoyle. It was better than dragging it around everywhere. The other trick was to stash it overnight along the river when I knew I would be returning to that section in the coming days. This saved me from having to carry it for hundreds of yards over rough terrain. I contemplated buying a couple of dozen stepladders and having them stashed at all of the prime barbel spots throughout the Korean Peninsula. One early morning at the Nam River tailwater, I went to fetch my ladder from a patch of forsythia, and it was gone. I looked up and down the riverbank for a quarter mile in each direction and did not find it. I never replaced the ladder.

DULL RED KNIVES

The sun crept up on the horizon, and a skein of mist drifted over the Geum River like a ghost passing through a graveyard. I had arrived after dark the night before and pitched my tent under an October full moon. I made a small fire, cracked open an OB Lager, and ate rice rolls smeared with wasabi and dipped in sesame oil. An hour earlier I had driven through a police checkpoint and had been asked to breathe into a small paper cup. I did so and the cop sniffed it. They were cracking down on drunk drivers.

It was the first day of tarpon camp. Three days of fishing and no obligations. Nothing scheduled. No family matters with the wife or down-island in-laws. No social obligations. In the past spring and summer, I had forgotten the news. I would teach a few classes a week, file a freelance story, and head back out. I was sleep deprived and sunburned and churning through cash. The fuel and tollway expenses were stacking up. I guided fly-fishing trips for trout but then went right back to the barbels and tarpon.

Usually, I camped near the river during these road trips. Other times— often during foul weather—I crashed at love hotels that cost around $30–40 per night. Back when Seoul hosted the 1988 Olympics, thousands of three- and four- and five-story hotels were built to handle the overflow of visitors. Before then, travelers in South Korea had two choices for accommodation: a handful of business-class hotels scattered throughout Korea's main cities and, if going deep into the rural areas, *minbaks,* or guest-stays at country houses where one shared the squat-toilet bathroom and slept on the floor in a side room. After the Olympics, these leftover small hotels with plenty of vacancies catered to Korea's massive sex industry, prostitutes and pimps, and to adulterers, secret lovers, sex fiends, and the underground homosexual population. The hotel owners responded likewise with condom machines in the hallways, soft porn tapes on shelves in the foyer, red light bulbs, pink linens, and the ever-present stink of cockroach spray, cheap aftershave, and women's shampoo.

Small-town love hotels were usually a cross between a Stalinist com-bloc and a gaudy whorehouse. Korean kitsch was always on display, and some were bedizened with minarets or neon coconut trees. They were places for lovers to discreetly screw, but they also made perfect flophouses for the angling flaneur. I grew to appreciate them as a back-up accommodation. The warmth of the *ondol*-heated floors, a hot shower, and the cozy quilts were what I needed after a rainy day of wading and fishing in waist-deep cold water. I learned to choose places where business seemed slow as there were fewer distractions. At one place, the moans and panting of a woman getting very well fucked in the next-door room were so loud and tormenting that I could not bear it. I cut my trip short and drove straight home to my wife.

I thought of all the desk work left unfinished in my home office. I was not troubled by this. It was time to disappear for a few days. Just drop off the map and forget to recharge the cell phone on purpose. It reminded me of words by Kwon Ho-mun, a sixteenth-century poet who ditched out of society:

> I lie on the bank of a river
> watching the water flow by.
> Time flows like that: a hundred years slip by
> in a moment.
> The ambition
> of ten years melts away like ice.

I was camped for river tarpon on a sandbar, and to get to it you take a back road and another backroad and a dirt road up a low valley, park behind a screen of pine trees for privacy, scramble down a steep bank, and then you arrive at my campfire. The coffee hissed atop the fire grate. Leftover kindling from last night's fire fueled the morning cook fire. *Samgyupsal*, the thick-sliced pork bellies, sizzled crisp in the skillet and I cracked a couple of brown eggs off to the side. From the chuck box, I plucked out a small container and scooped a dollop of homemade refried beans into the mix.

The fog on the river immersed me in a bone-penetrating cold, and I shifted the cook grate and tossed more wood on the open side of the firepit. I took care of my chores and broke down my tent so that when the coffee and breakfast were ready, I could enjoy the morning. There is anticipation of making camp and getting gear organized just so, but there is also a satisfaction to breaking camp, moving on with the day ahead.

This October trip would be the last of the year. The river tarpon get active in the spring and tend to fish well even in the hot months of summer, and then the fishing declines slowly as the autumn cold sets in. The phenological indicator for when the tarpon start moving is the season of yellow dust that occurs in late March and into April. Annual winds blow over China's deforested and greed-wrecked lands and carry sand particles into the Korean skies. The polluted dust forms deranged electrified sunsets as if the Yonbyon nuclear reactors in North Korea just blew a gasket.

I grabbed my cruiser axe and went over to a stand of pines and found what I was looking for—a rotted trunk of a fallen pine. I knocked away the spongey tissue, and all that was left were the petrified knots and the hard core of fatwood. I hauled the chunks over to some rocks and set them there to dry in the sun. They would be the evening entertainment in the firepit. I would most likely return here to camp for the night, but my gear would come with me.

As I sipped on my coffee, I heard a whistling sound through the mist. It got closer and I could hear them. I stood up and looked downstream. Their silhouettes broke through the fog, and they appeared for a couple of seconds before they disappeared back into the vapor. Baikal teal, *Anas formosa*. They were migrating from their breeding grounds in the great taiga oblasts of Russia's Far East. They winter in South Korea, Japan, and China.

The drake is the most beautiful of all the teal species, and it rivals the mandarin duck, the harlequin duck, and the North American wood duck in its elegant colors and plumage. The huge population is an accident. The stubble in harvested rice fields adjacent to man-made reservoirs (which destroyed natural wetlands and dammed up rivers) made for good feeding areas, and the artificial lakes were safe roosting spots. A year before, at nearby Seosan, I had watched a flock of around 20,000 Baikal teal take flight in the evening, and their synchronized swarm of swirling flight patterns had blotted out the sunset.

The Baikal teal might have lucked out, but all of the wildlife in the area is in peril. South of the Geum River estuary is the Saemangeum Tourist Information Center, a monument of government propaganda that green-washes the reclamation of the tidal flats. It is located on a hill that overlooks the ongoing development. Inside are displays claiming the reclaimed land will be turned into a Potemkin eco-paradise. A short distance down the road is the local residents' answer to the information center, and the sight is eerie. Out in the tidal flats are *jangseung* totem poles planted in the mud, along with a wooden scaffold that hoists a rotted fishing boat. There are also *sotdae*, another kind of totem that is a wooden pole topped with a carved bird or fish. Tied to the poles are ribbons and banners protesting the development project.

This region will be forever changed by the Saemangeum reclamation project. Building the world's largest embankment seawall, twenty-one miles long, will eventually cause the tidal flats to be turned into landfill that will be used for golf courses, rice fields, and various industries. It is one of the world's greatest shorebird resting and feeding sites and an essential site for the endangered spoon-billed sandpiper. Environmental groups have rallied hard against this, but Korea's construction cartel finally won the court battle when the Seoul High Court ruled, in December 2005, that the cement pouring must continue.

I took the fly-bys of the Baikal teal to be a good omen. It was the first day of tarpon camp, but it could be the last day of tarpon camp. The next two days would determine if I would ever purposefully pursue this species of fish again. I would always appreciate catching river tarpon, but they were shaping up to be an underwhelming game fish—totally unlike its saltwater doppelgänger. They were nearly identical in shape and body structure. These freshwater species were much smaller, and that is perfect with light tackle. The problem was that the Korean river tarpon did not have the fighting traits of the famed saltwater tarpon: no epic leaps, no blazing runs, no forearm-burning battles. Instead, it struck the fly aggressively and fought for a few seconds, and then it flopped over and gave up.

It was a strange and disappointing behavior. At first, I thought it was a fluke, but after catching enough of them, I noticed the pattern. It was the same quick strike, a flutter of initial fight, and then it went limp as it was easily brought in—like reeling in a twitching dish towel. My only theory was that larger fish might have more vigor. They can grow up to forty inches, but most are in the twelve-to-twenty-inch range. The ones I had been catching were in the upper teens, and I wanted to at least break the twenty-inch or thirty-inch mark and see if larger specimens could exert more savagery when hooked.

I systematically fished and explored new spots over the peninsula, often doubling or tripling up on marking species, as in hitting a large section of river looking for half-beards and tarpon and then following an upstream tributary deep into a mountain valley with the hope that a relic population of trout might exist. There were many, many fishless dead ends.

For tarpon, I fished Chungju Lake near Hwa-am Village and the entire length of the Jaecheon River, which flows into it, the Namhan River near Danyang, and the Dalcheon River and revisited many of my old bass fishing spots on the Nakdong around the Pusan Perimeter. I got to see a lot of good country, but a lot of bad country too.

I spent a few days fishing and drinking at the Three Rivers Tavern (*samgang jumak*) near the city of Yecheon. It is tucked along the convergence of the Naseong and Geum streams and the Nakdong River at the village of Samgang (hence the name, Three Rivers). In the old days, ferryboats and salt barges cruised the waters of the Nakdong, and boatmen would stop by the pub for a few cups. Constructed around 1900, it was recently designated as a local folk asset and earmarked for restoration. It's not exactly a tavern or pub in the Western sense, but more of a semi-deserted ramshackle drinking house with a good view from the porch.

Nearby was Hwaryong-po, a horseshoe bend in the river that creates an enormous sand beach and is encircled by mountainsides. I climbed up one of the mountains to visit Jangan-sa, a thousand-year-old Buddhist temple. The next day, I fished early in the morning and met some carp fishermen who eyed my fly rod as if I were holding a Buck Rogers space gun. They invited me to watch some cockfights, and we spent the afternoon gambling and quaffing *makeolli*, a crude rice wine.

The Geum River was the last on my list. If I did not catch any sizable river tarpon here, I would table the species—meaning, I would not drive any distance to catch them. I would drop my research and field studies on them, and I would not offer them as a species of pursuit for clients. They would merely be an interesting bycatch. I had already added fly fishing for half-beards as part of my guide service, and the angling experience for clients was always a success.

As I scouted half-beard water, I also shopped around for tarpon. The Holy Grail was to find a shallow-water flat loaded with barbels and have a nearby section of deeper water filled with tarpon. Instead, I found good spots for half-beard and nearby spots where tarpon lurked but were unpredictable in their appearance. Sometimes they were there, other times not. I wanted more consistency. In the past few months, I had been endlessly scanning maps about the distribution of tarpon, once thought to exist only in rivers flowing into the Yellow Sea but now found in numerous inland watersheds.

Like many fish in Asia, *Erythroculter erythropterus* has an odd mix of names. River tarpon is one. It is often referred to as "skygazer" because of its eye position. It is a predatory carp that ambushes minnows, in the upper part of the water column, from below. It is always looking up. This is a tell on what kind of flies to use: nothing that sinks much.

While studying Korean literature about the fish, I found that *skygazer* is sometimes spelled as *skygager*. This is because there is no z sound in the Korean language. It's a common misspelling. The closest thing to the z sound is a j or g sound. As an English teacher, I taught many lessons in how to pronounce *zookeeper* instead of *Jew keeper*. *Zack* instead of *Jack*. *Zap* instead of *Jap*. The best I've come up with as a teaching heuristic is to have them imagine the sound of electricity and make a buzzing vibration sound in their mouths that sputters out with a z-like sound. If that doesn't work, I have them mimic the sound of a mosquito (*mogi*) buzzing in their ear.

The skygazer is the only member of the genus *Erythroculter* and is described as *Culter erythropterus*. This is created from the Greek word *erythros*, meaning "red," and the Latin word *culter*, meaning "knife." Its Korean name is *gang-joon-chi*. Other subspecies are called the Chinese red-sided culter, redfin culter, and topmouth culter. In Taiwan, legend has it that Chiang Kai-shek liked eating it, and it got the nickname the "president's fish." It is found in China, the Russian Far East, Mongolia, and the Korean Peninsula.

Wading is necessary, for the extra reach to hit the right presentation in the water column and also to cover the largest area. As the water column warms, the tarpon cruise near the surface to inhale minnows that stray from the safety of inshore cover. This was mostly fishing blind—as opposed to sighting half-beards from a distance and being able to watch the entire casting and catch sequence before your eyes. If barbels were death from above (feeding downward on the riverbed), then tarpon were death from below—eyes positioned atop their heads and ever-searching for prey they could ambush from under and behind.

I walked down the rice paddy access road for a couple of miles upstream. The plan was to fish my way back to the car and regroup. The riverside and surrounding area was the standard mélange of rural Korean lowlands. It was where more human activity happened. Rice paddies patchworked off into the distance and were punctuated by powerlines. Small clusters of houses with their own courtyards formed their own small villages. Off in the distance I heard a familiar *ppongjjak* song. It is a genre of music that many

Fig. 10. The Korean river tarpon, *Chanodichthys erythropterus*, is a sleek, aggressive hunter but a very weak fighter once it is hooked. Photo by James Card.

Korean old timers love. It is a two-beat rhythm that forms a primitive pop, folk, and lounge-music sound. For a long time, I thought this music was very lame, but it grew on me and I developed a sentimental fondness for it.

I hiked past a miserable-looking Jindo dog on a short chain. He barked and I muttered back. I walked by a warren of decrepit houses. On the global stage, the government was pushing its very first nation-branding agenda, "Dynamic Korea!" with images of lots of high-tech innovations and hip refinements, but that is pretty much a lie once you get out of the lusty neon-lit cities. There are no young people in rural Korea. On one building was a banner advertising mail-order Vietnamese virgins available for marriage. South Asian women watch Korean soap operas and imagine the country to be advanced and modern and hip and figure marrying a Korean bachelor might be the ticket to a better life. Then they end up in a place like this and lose their minds. It is the old bait-and-switch.

Like most Korean rivers, the Guem River is embanked with cement and riprap, lending it an artificial trench-like appearance but with enough room to meander and form small sandbars and braided currents. I noted the debris left behind by carp squatters, a mix of *soju* bottles and *ramyeon* noodle packets. Weirs and dams are staggered throughout its entire length and have formed pools.

I cast out a muddler minnow. The red knives are inhalers and have mouths like a white crappie whose nickname is "papermouth." Its mouth is of a thin, translucent cartilage that hinges tight when closed but when agape, the maw is surprisingly wide, for sucking down minnows. It is sleek and silver, and its mouth is a vacuum snout for inhaling small fish at high speed. A school of small tarpon will surge at any movement and rush at any animated fly.

My first perfunctory cast elicited the first strike of the day. I set the hook, and my fly line snapped tight like chain lightning. In the first second I knew it was a nice tarpon. It boiled on the surface and then went limp. I reeled it in, unhooked it, and released it. It was a good start. Where there is one, there are more. I caught another and another, working along the river-left bank so my right-handed casts could have clear backcasts over the open water and to make down-and-across swings. Each cast and each swing were two footsteps to comprehensively cover the most water: cast, swing, two steps, repeat.

The Guem River is the third longest river in the country, and it flows into the Yellow Sea at the port city of Gunsan. An American airbase is located there and is the home of the Eighth Fighter Wing, known as the Wolf Pack, the fliers of F-16s. It was along this coast in 1904 that a twenty-eight-year-old Jack London sailed in a hired fishing boat to Chemulpo, the old name for the city of Incheon, the same port where MacArthur launched an aggressive amphibious counterattack to turn the tide of the Korean War. London covered the Russo-Japanese War for Hearst newspapers and used rice paper to take notes. He was arrested by Japanese authorities after beating up a man for stealing feed from his horse. President Theodore Roosevelt brokered his release on the condition he leave Korea immediately. London wrote more dispatches than any other journalist covering the war.

Nowadays the river is called the Geum River, meaning "gold river," likely derived from the tint of muddy sand. It was called the White Horse River from a myth of the ancient Baekje Kingdom involving a gleaming white horse. It takes some time to ponder, but among the few foreign visitors to have hiked the sandy shoals of this river are a few anglers like me and a handful of international birdwatchers. The only other Westerners that have walked on these sandy banks are those that served in the early days of the Korean War. In mid-July of 1950, American forces formed a line along the southern bank of the Guem River to make a last-ditch effort to hold back the descending Communist forces. The North Koreans crossed by makeshift rafts, waded neck-deep, or just swam across. The enemy fire to cover them was so powerful that General Meloy compared it to the European theater of World War II. They were forced to withdraw, and one out of every five American soldiers died in battle.

As I wrote down a few notes in my field journal, I left my fly line hanging in the water. The current caught it and slowly straightened it out. No big deal. The drag on my reel was set light as these tarpon are such weak fighters they never took the line to the reel, and I never expected them to. The current tugged on the outstretched fly line, and I heard the ticking of the drag as more line peeled off the reel. I kept writing. When I pocketed my journal, I looked down at the reel and the backing was exposed. Almost the entire length of the fly line—ninety feet—was extended downstream. I reeled in the line and then I felt the weight.

Something was on there, and it wasn't a snag or a stick. I could see a small wake draw nearer as I cranked in the line. I felt a few quivery vibrations. As it came closer I could see it was a fish, another tarpon, but it was the largest of the day—twenty-one inches—and a personal record for me. I never felt the strike, just the slow tug of the current pulling the fly line downstream. It took me about five minutes to jot down those notes, so the fish could have been hooked for a couple of minutes without me knowing it.

It is an amusing angling curiosity to catch a fish "on the dangle." I always instruct clients to leave a dangle. A dangle left out is when you stop actively casting. Perhaps you put on some lip balm, drink some water, eat an apple, or bullshit with your fishing partner. Usually at this moment your fly rod is tucked under your elbow or against your shoulder. I have a couple of tabs on my fly vest that hold the rod against my chest so my hands are free for other things. The dangle is simply leaving your fly in the water, just dragging in the current. The length of line out doesn't matter. I've seen fish caught with ten feet of fly line hanging off the rod, and other times when his fly line is stretched halfway downriver. When the angler lifts his rod, there is a moment of dumb surprise: I think I got something.

The dangle can get you in trouble too. It can get snagged, and it will be annoying to hike over and untangle the mess. Also, the fly doesn't matter much when on the dangle. Obviously, streamers and wet flies would be the best for this passive non-presentation. A dry fly will be submerged and sodden, but sometimes it will still pull in a fish, especially if the dry fly also has a streamer-like shape, such as the Hornberg. There are certain conditions where a dry fly might get sucked down over a waterfall and into the next pool. That is bad. But if the fly line happens to be the right length—within inches—I have seen the fly bounce on the water at the edge of the plunge, and it behaves like a skittering bug, which sometimes attracts the attention of the loner trout that hangs at the very end of the pool.

The dangle is an illustration of why some anglers catch more fish than others: their fly has more time in the water. Every moment on the water is maximized, even when taking a break and letting your fly hang in the water. Stand behind me with a stopwatch, and as soon as my fly hits the water, the stopwatch is activated. Then when my fly comes out of the water, the stopwatch is paused. If you were to do that the entire day, it would reveal that my fly spent more time on or in the water than yours, perhaps double or triple or quadruple. Time is wasted by not fishing effectively and efficiently. Time is wasted by snags, which means poor line control. Time is wasted by retrieving a fly from the bushes, which is negligent backcasting. Time is wasted changing flies when it really doesn't make a difference what the hell you have tied on. Time is wasted by navigating upstream and not taking the smartest route to the next casting position. There is nothing wrong with wasting time on a trout stream or river, but it should be a pleasant way of wasting time, such as daydreaming at a small waterfall, rather than untying a messy knot.

Most of these headaches are caused by slack in the fly line. Slack is the root cause of all problems in fly fishing and life. To be called a slacker is to be a person known for shirking obligations and necessary work. Your obligation as a fly angler is to always have complete awareness of the slack in your fly line. This includes the fly line that has been stripped in and is coiled by your feet, and it includes the length of line that stretches from the rod tip to the fly in the water, many feet away. This is your obligation. This is necessary to catch fish.

So the large tarpon came from the water farther downstream. I skipped over the nearby flatwater I was working and hiked down, and I saw the tarpon come from a handsome pool, deep and with a medium-flowing current. I caught another nice one, about the same as the one caught on the dangle. It smashed the fly, and once it felt the line, it splashed about hard with a body-twisting rattle, and then it went limp with just a few splashy spasms as I brought it to net. I repeated this with two more tarpon. Then I caught my largest ever, one about twenty-four inches long. It attacked the fly aggressively, fought for a couple of seconds, and went waterlogged.

I hiked back to the car for an egg-salad sandwich and a cup of coffee. Egg-salad sandwiches had become an oddity in my life as a fly-fishing guide. I was scrounging in the fridge one day for something to eat, and nothing looked appealing. And then it came to me: make an egg-salad sandwich. It had been a long time since I'd had an egg-salad sandwich. I made one, and it was simple and delicious and satisfying.

I provided lunch and snacks for my clients, and the bitch of that was figuring out a menu that would make everyone happy yet be of ingredients that were easy to round up. A decent ham-and-cheese sandwich was difficult. Korean ham was one step above SPAM, and Korean cheese was nothing more than solidified milk, vegetable oil, and yellow food coloring. I could serve all kinds of wonderful Korean foods, but I felt it was not my place to introduce exotic Korean dishes to the inexperienced palates of my Western clients. The goal was to always keep it simple, tasty, and nourishing. Egg-salad sandwiches were the answer. They became a nostalgic wonder, and the clients always made the same three comments:

"Man, I haven't had one of these in years."

"I haven't had this since I was a kid."

"You have another?"

I always made extra. Nobody went hungry.

I would not be serving egg-salad sandwiches on guided fly-fishing trips for tarpon because there would be no more purposeful fishing for tarpon. My decision was made. Tarpon camp was over early. There was no more investigating to do. Korean river tarpon were now buried in the Boot Hill of my mind. I had set out many moons ago to determine whether two doppelgängers could be the next great freshwater game fish. I had learned that one offered a spectacular angling experience and the other was mediocre. I packed up my gear and drove to the Jiri Mountains.

7   Bear Medicine

I arrived in darkness and was on the water by nautical twilight. I suspected the pools and riffles were not fishing as they should have been. Although I put my clients on cherry trout throughout the spring, the number of strikes was off. On runs where an angler was almost guaranteed a strike if the fly was well-placed and drift-free, there was nothing. No strike on the glassy surface. Not a shot of fish silver coming up from the depths to even look. I watched the sun come up while sipping coffee out of a tin cup. I ate an apple and then unpacked my diving gear.

In South Korea, summer is more than a high season for tourism. The school and university students are out on break, but also the large corporations grant their employees leave at almost the exact same time, creating a massive exodus to the countryside, beaches, and scenic valleys, and the Jiri Mountains are no exception. I avoided the Jiri Mountains throughout the summer. Fishing was impossible during the tourist season, when nearly the entire country takes their annual vacation in lockstep. Lawn chairs and picnic tables were often dumped into the trout pools for splashy playtime furniture.

I kept guiding trips but warned experienced clients of this phenomenon and steered them toward half-beards, largemouth bass, notchmouths, and smashmouth perch. With beginning fly anglers, we spent mornings on flat water to practice casting and to catch huge bluegills with foam spider flies. In the afternoons, we fished fast water for rainbow chubs. These were mountain streams just as beautiful as the trout streams I fished, but they held only these playful fish and a few rare keokji. By their common name they are referred to as "pale chubs," but the naturalist that gave *Zacco platypus* that moniker must have been color blind. Instead, I called them "rainbow chubs" because of its shimmering peach fins, amber and indigo sides, and iridescent green back. It has wart-like tubercules around its snout, and they grow to be around ten inches long. They were perfect for beginners because they took every dry fly thrown at them with abandon.

Although the Jiri Mountains are topped with snow in the winter, the winter in Korea is a dry one with low precipitation, and the water still flows, albeit at a reduced volume. Throughout fall, winter, and spring, I had these Jiri streams almost entirely to myself, but I avoided them during the tourist season. I drove up once in late July to check out the scene. I parked on the highway bridge and gazed down the valley to see hundreds of picnickers covering the rocks—swarming and splashing in the pools and squatting on the rocks cooking lunch. A few crouched on the boulders with simple cane poles. A couple of others were throwing cast nets into the water and dropping inverted cone fish traps into the stream. I drove farther up the valley, and the road was swarming with cars and busloads of tourists. The quiet village howled with their drunken hoots and garbled karaoke.

For the rest of the summer, I had trouble sleeping. I dreamed of empty pools devoid of trout on my favorite trout streams in the Jiri Mountains. I worried that poachers might be overrunning the streams. Tourists came every summer with cane poles and nets, and they would pick off some trout. I thought the damage would be minimal as most were incompetent, but it concerned me. I wondered if there was a hardcore crew at work, systematically clearing out the trout, pool by pool. In South Korea, one can catch and kill as many fish as one desires by any means possible. Catch and release is a foreign concept. As the only fly-fishing guide in the country, it was something I had to tolerate. But I refused to tolerate it when I came across it. I destroyed unattended fish traps, trotlines, and gill nets wherever I found them, and more had been appearing in the Jirisan streams in the past year.

The poachers had active methods besides the passive fish trap techniques. One was a Korean version of the slaughter pole. In the American South, the slaughter pole is composed of a cane pole topped with a short line of mono and tied off with a jig, bait, or lure. The angler reaches the pole out

Fig. 11. The rainbow chub, *Zacco platypus*, is the perfect species for beginning fly anglers to understand dry fly presentation and hook set. Photo by James Card.

to a stump or other likely place a big bass would be lurking and smacks the lure on the surface. Naturally, the bass smashes the lure, and then the angler hoists in the fish. The Korean version is applied underwater in the mountain creeks. A bamboo pole is topped off with a few inches of fishing line, and tied to the line is a hook and worm. The angler then walks up to a pool and studies the crevices and low spots where a trout might be hiding. He jams the pole right down into the streambed near where the fish is taking cover under some rocks. That few inches of line dangles the hook and worm just off the bottom and right in front of the nose of the trout. Eventually it becomes too much, no matter how spooked the trout is and regardless of whether its overhead cone of vision detects a strange humanoid figure above the surface. It takes the bait; the angler needs only patience.

The picnic hordes were gone. I had the stream to myself again, and it left me feeling satisfied in a selfish way. For the rest of the year, I would see no one else on the stream except for the mushroom growers, beekeepers, orchardists, and maple sap harvesters; the odd Buddhist monk; the mountain spirit pilgrims; and the guesthouse keepers and part-time construction workers. All of them were old timers.

The Nayday was covered with hundreds of stone cairns built by urban tourists trying to add meaning to their empty lives. Litter was everywhere. I looked downstream and counted the carefully balanced stacks of stones. The trout pools turned into swimming holes, and garbage was always left behind in the most thoughtless manner: sit down for a picnic on a flat boulder overlooking the stream, eat lunch, and then get up and leave everything behind. I slipped a couple of garbage bags into my fly vest and jumped down into the boulders. Magpies and crows picked through the edible remains and squawked off upon my approach. I kicked over two cairns, the first of many. I did not want the trash washed downstream, so I collected it as I hiked along. I grabbed plastics, paper, and aluminum. I left hundreds of wooden chopsticks as they were at least biodegradable.

In Sten Bergman's book *In Korean Wilds and Villages,* he wrote about flooding in the Jirisan region: "During the night there was an appalling catastrophe. The river, which runs through the town, overflowed its banks and completely destroyed more than a hundred Korean houses. About seventy-five people were drowned that night. The force of the torrent was so great that it was very difficult for old people and children to be saved."

Every year I made it a tradition to drive up to the Jiri creeks during the typhoon and monsoon rains. There was no intention of fishing. It would have been suicidal. One misstep and you would be washed downstream and smashed into boulders like a whitewater crash-test dummy. The flooding rearranged boulders the size of dump trucks. Along the edge of the forest, I once came across football-sized rocks smash-wedged into the crotches of forked pine trees. It was another theory of mine, that perhaps the recent dearth of trout rises was not from poaching but from trout getting rearranged every year by the floodwaters.

This led me to consider how the trout migrate up such a steep gradient. This wasn't a matter of a salmon leaping over a small spillway. This would be a rocket launch to get to the next pool. With the steep gradient and waterfalls, I pondered how the trout moved up or down the river, if at all. Some waterfalls were too tall for a trout to get over, but there were also hidden side channels buried under the boulders that led up to the next pool, like a hidden plumbing system. I studied scientific papers about fish barrier analysis and the burst velocities of trout. Some trout simply got washed down into the next pool, and the ones that didn't had plenty of crevices at the bottom of the pool to hide under as the massive volume of whitewater raged above them. Some pools got cleared out, while the populations of other pools remained stable because of better cover. They would then live in that pool and eventually repopulate it.

This enigma relates to the concept of source-sink dynamics, a model ecologists use to figure out how organisms repopulate or depopulate patches of habitat. Creatures might live in a good habitat and can reproduce better (the source), while others live in a poor habitat and do not reproduce as well, if at all (the sink). If creatures from the source meander into the sink, that population can continue to exist. In the case of the steep-gradient boulder streams of Jirisan, it isn't so much about habitat quality (although upstream the water is always cleaner and colder) as it is about the fact that the network of pools in the creeks is always fluctuating, year by year, due to forced migration (trout getting washed into lower pools) or lack of migration (trout unable to swim or leap into the upper pools).

The creek is a population of isolated groups connected by flowing water. The trout that populate the individual pools live independent of the trout in the other pools. The smaller the population in a pool, the more likely the trout in that pool will become extinct until other trout get washed into it and recolonize it. Nature abhors a vacuum, and everything—from the big pools to the tiniest of pocketwater—gets repopulated after a major flood event. Therefore, I caught and released every trout unless it was mortally injured.

Source-sink dynamics also applies to South Korea's rural population. Everyone in the country is old. The birth of a child in a village is a rare event. For decades, the younger generations moved to the cities (good habitat, opportunity, the source) and have not repopulated the rural areas (poor habitat, no opportunity, the sink). The few people who do move back to the country tend to be retired or semi-retired, and they open guesthouses as golden-years lifestyle businesses. The populations of aged rural villagers and trout in isolated mountain streams are sink metaphors for each other.

I was intrigued by the idea of learning more about where trout take cover in the far depths. There were times when I approached a pool while daydreaming and watched the trout rocket to the deep, not to come back up for a long time. I always wondered where they hid and how many trout

I didn't see. I remembered an angler in New Zealand telling me fish counts were conducted there by scuba divers making a line across the river and floating downstream and marking trout. I guided many American military personnel stationed in Korea, from privates all the way up to a four-star general. I remember one captain telling me about his time in Iraq. He mentioned the phrase "ground truthing," which was a term to describe the reality of what was happening on the battlefront as opposed to the sketchy intelligence reports that filtered up the chain of command. That's what I had to do, water truth the trout stream and see what was really happening in the depths of the pools. Above the surface, I was getting only part of the picture.

Behind some boulders I changed into my wet suit from my spearfishing days. There was a gaping hole in the backside of the wet suit. When spearfishing the coastline, I took rest breaks on rocks covered with sharp barnacles, and this, in turn, eventually shredded the neoprene covering my ass. While putting it on, I felt like the gimp from *Pulp Fiction*. I squatted in some riffles to let the icy water ooze into the ratty bunghole, and then I submerged myself so the water would infiltrate the rest of the wet suit and my body heat would warm the water trapped between my skin and the neoprene.

With the wet suit slicked, I rock hopped up the creek with my backpack. I memorized the pools and runs and gave them names: Four Pools, Big Bend, Radio Run, the Rotten Cement Bridge, Bamboo Point, and Prayer Pool. Smaller patches of riffles had no names. I would just recognize the familiar water and say to myself, "Ah, this one. I know this one."

Four Pools was the one I wanted to study first. It was one big pool disjointed by tumbled boulders separating it into four distinct sections, each holding separate trout in separate holding patterns. I shimmied over a rock and stretched my neck toward the water until my mask touched the surface. The trout were there. They were beautiful and wild, and the water was so clear I could see forty feet across the pool. There were smaller ones, six-to-eight inchers grouped together, and a couple that neared a foot in length cruised the lower depths. I wanted a closer look, and I slipped into the tail end of the pool like an enormous mutant otter.

Each of the four sections held trout. They lived among the underwater boulders, which offered hundreds of nooks and crevices that provided

Fig. 12. Steep drops form the boulder garden pool–waterfall sequence in Jiri Mountain streams. Photo by James Card.

protection from above and relief from the current. The routine for the rest of the day went like this: scamper up some boulders to the next pool, arrange my body in the rushing tail waters in whatever yoga position was required so I could have a direct line of sight straight up through the pool. From there I spied on the trout without them spooking into their hiding spots. After some observation, I took a gulp of air through the snorkel and pushed off into the pool, scattering the trout. I counted trout, but another objective was to survey the underwater dungeon and see how the currents bent around and through the rocks and crevices.

I kept a tally on a Rite-in-the-Rain notepad after every swim. In some pools I was right: there weren't as many trout as I had expected. Others were of average density, and a few surprised me with schools of juvenile cherry trout stacked up in the main feeding lanes.

The pool count:

Swirled Riffles held three cherry trout.

Elongated held four trout.

Prayer held nine with two around fourteen inches.

Fossil Rock had over twenty cherry trout with two lunkers that held tight in a deep crevice.

Corner Pocket held thirty juvenile cherry trout.

Balanced Boulder held eight cherry trout and a huge number of chubs—the most I had ever observed advancing this far upstream.

The Double Waterfall held six cherry trout, facing each other on opposite sides of each waterfall.

Darkwater held two cherry trout, a few chubs, and one super chub.

Precursor held only one cherry trout. To note, a footpath from a guesthouse led directly to this pool. Easy fishing access.

Far Side held thirty-four cherry trout with one lunker.

Boulder Overhang held seven cherry trout with a small one being chased by a brook perch. This was the first sighting of a brook perch so far upstream in the river continuum and at such a high elevation. Also, under the overhang was a paper wasp nest the size of a beachball that I had never seen before.

Radio held eight cherry trout with two nice ones, and one of those was the largest trout of the day.

S-Bend held zero trout.

Big Bend held eighteen trout, medium sized.

Horseshoe held one little trout and one big one.

Four Pools 1 held fourteen cherry trout and a few rainbow chubs.

Four Pools 2 held twenty-five cherry trout with one lunker and a mix of chubs.

Four Pools 3 held twenty cherry trout.

Four Pools 4 held three cherry trout and one Chinese softshell turtle.

Rotten Cement Bridge held over twenty cherry trout, with three of them lunkers, some chubs, and one curious brook perch,

I was excited about the brook perch. I did not know they existed so far up into the mountain headwaters, but I was not surprised that I had never caught one. They are bottom-up ambushers and would never bother with dry flies.

One pool gave me the biggest revelation. The biggest cherry trout I had ever caught on this stream was around fifteen inches. Any trout that is a foot long on this stream is a trophy. While fishing the Radio Pool a few years ago, I spotted the biggest cherry trout I had ever seen on this creek. I could see only its head, gills, and a sliver of its dorsal fin. Judging from that portion of body mass, it must have been close to eighteen or twenty inches. I moved in to get a closer look, and it edged back, disappearing into a dark hole.

For the next three years, I tried to catch this fish. I directed every client I guided on the Nayday to cast near that hole. I tied on big meaty streamers, big nymphs, little nymphs, and all the freak flies that I owned. The fish never hit the flies, nor was it ever seen again.

Cherry trout are fall spawners, and their bright colors turn a darkish brown during this time. I made a solo trip up the Nayday when the mountains were blazing with autumn colors. I worked my way up the creek and came to the pool that held the big cherry trout. There he was, exposed in some shallow pocket water. His body was darkened and covered with a pale scum. I stepped closer and the fish didn't move. I poked him with the tip of my fly rod. He twitched and finned forward a few inches. The fish was sick and dying. Of what I did not know. My mind raced. Old age? Perhaps. I thought of all of the trout diseases I had read about: whirling disease, gill lice, and many others, and recalled reading an abstract about *Oncorhynchus masou* virus disease. Then I thought of the possibility of a disease wiping out the entire population of cherry trout on the Nayday.

I grabbed the trout by its tail and flipped it onto a flat rock. It did not fight or squirm. Its gills barely moved. I slid the point of my bird and trout knife into its head and punctured the brain and put the fish out of its misery. I rinsed my hands and took some photos of the fish. I tossed it into a bamboo thicket so if it did harbor a contagious disease, its carcass would not contaminate the water.

I sat for a while thinking about this odd turn and worried that some contagion could threaten these fish that I so dearly loved. My stomach growled, and I realized I had not eaten anything all day. My work was done, and then I remembered that while driving in I had seen a sign for the bullfights. The bullfighting stadium was not far from the highway that I used to drive home or to drop off clients at the bus station, their hotels, or the Jinju airport. It was an easy detour. When I neared the intersection, I would ask if they wanted to go see a bullfight. The answer was always yes.

I packed up and changed, and I drove down to a *bulgogi* house in the foothills that I sometimes frequented. Near the doorway an old crone whispered mantras and burned a mugwort incense coil that smelled of citronella, pine, and marijuana. I ordered a Korean version of carpaccio called *yookhway*. The service was always fast since the owner didn't have to cook it and she knew that I ate alone. I ordered and opened a bottle of Gamgrin, a persimmon wine that is aged in an abandoned train tunnel near Cheongdo. The *yookhway* is similar to steak tartare, but the lean sirloin is cut into thin strips and served chilled so the beef melts and warms in your mouth at the same time. There is the taste of soy sauce, sesame oil, garlic, and pepper. The beef is topped with a raw egg sprinkled with chopped pine nuts, and it is presented on a bed of julienned pear. I finished off the plate and slipped the bottle of persimmon wine into my pocket and left for the stadium.

Unlike the Spanish bullfights, or the bloodless Provençal and Portuguese bullfights, the Korean version is not man against beast, but bull against bull. The sport originated as a farmer's pastime and a way to earn grazing rights. The stronger bulls were later selected for breeding. The National Assembly passed a law legalizing betting on the bullfights, and Koreans love to gamble. The stadium is always packed with hustlers who study each bull like other handicappers study racehorses.

Outside the stadium a lady sold traditional pumpkin candy called *hobakyut,* and others offered paper cups of steaming silkworm larvae, roasted ginkgo nuts, and scraggly corn on the cob. A Buddhist monk in a pith helmet clacked a wooden gong signaling for alms. An elderly woman carved off a sliver of arrowroot and offered me some. Near my feet, a beggar with amputated leg stumps wrapped in inner-tube rubber dragged his body on a roller board while pumping an accordion.

I found a seat as a match finished with a cowardly bull backing off and refusing to engage its opponent, who stomped and snorted. The winner was determined by the bull who gave the best fight. The bullring was a fenced corral and covered with sand, and a judging platform overlooked the scene. The bulls were judged on their attacks: a hook to the neck (*mokchigi*), the flanking push to the side (*yopchigi*), pressing against each other's heads (*tolchigi*), horns clashing with a head attack (*yonta*), and other offensive maneuvers. The bulls are in fighting shape as the owners run them on sandbars and mountain trails and have them pull tires and head butt tree trunks. They are raised on a mass-building diet of hay, beans, and barley.

Near competition time, they are fed eels, mudfish, snakes, and live octopus as a stamina nostrum.

Sometimes there was blood when a stray horn gored a flank, but that was uncommon. The next farmers brought the bulls into the circle, a caramel brown one—a common color—and a chocolate bull with a rusty streak mottled down his back. Their horns were about a foot long and polished but dull pointed. The bulls did not want to fight. They were mellow and content. The farmers and referees pushed and prodded the beasts toward each other. One farmer leaned his shoulder into the bull's giant rib cage and shoved forward. Another man twisted the bull by its horns to point its head in the direction of the other bull. The bulls snorted and wheezed a few times.

They caught each other's scent and a change occurred. The men backed off a few steps and the muscles of the bulls quivered, and their heads swung and bobbed. The instinct of the male beast took over: the urge to fight for territory and hierarchy and procreation dominance. They pawed the earth, splashing sand up on their flanks and backs. They sniffed at each other's skulls for a couple of minutes and, in an instant, they smashed heads and knocked horns. It was a pushing match of power, balance, and position like a sumo fight. The two bulls raged head-to-head until the chocolate angled and hooked a horn into the other's neck. The caramel countered and butted into its shoulder. The handlers hop-footed near their hindquarters and tried not to get trampled.

The chocolate broke off and spun away bucking. The caramel stood quietly as if in a peaceful pasture far away. The bucking bull stopped and bellowed and pawed the ground. They sniffed at each other again, and they clashed for a few more minutes, sand spraying everywhere. They paused, flanks lathered and foaming at the mouths. Nostril slime dripped off their brass nose rings. The caramel swung his neck and charged, and chocolate answered with the hard, hollow sound of skull bone rung like a bell. They locked horns, necks shivering with strain and noses nearly touching the sand. The chocolate slid his head under, gained leverage, and pushed into the caramel's shoulder, knocking him off balance. The crowd cheered. The caramel regained his footing by skidding around and squaring up. The chocolate charged, caught the caramel under the neck, and pushed until he broke away, giving a grotesque bellow and running a half circle around the ring. The caramel refused to fight after a few minutes. The judges announced the chocolate bull as the winner. The owner would collect prize money, and the spectators spun in their seats to settle and collect debts and to lay down bets for the next match. I watched a few more matches and then drove home to study trout diseases.

A few days later, I contacted Dr. Mamoru Yoshimizu, a professor of aquatic epizootiology at Haikkaido University. I emailed him photos of the big cherry trout and hoped he could determine the cause of death. He kindly replied that after the spawning season, mature *O. masou* salmonids often come down with a fungal disease. It is common he said.

I did not feel bad for the big trout. He had a long life and was smart enough to dodge the slaughter poles of poachers, along with refusing everything in my fly box. He lived in a secret world, and if it weren't for my water-truthing mission, I would have never known of his existence until I found him half dead. That would have been a shame. Catching a large fish is good, but what is better is the anticipation of knowing they are out there.

The next weekend I water-truthed Jungsan Valley. A sign near the bridge warned of flash floods and reported deaths in the past. I parked and hiked past a shed with rafters of hanging persimmon left to dry and went down a slope covered with pumpkin vines and beehives made of hollowed logs. This valley gets hit harder than the Gollum Valley, and the trash left by summer tourists was everywhere. I repeated my routine: bag up trash at each pool, dive it, count fish, record notes.

Farther upstream was the Graffiti Pool, where a few boulders are hand-painted with Hangul letters. A client once asked what they meant. I told him it said, "No Fishing." He gave me a nervous look. He was an international banker with a reputation to consider. "Relax," I told him. We were fishing outside of the national park. "It says 'Minsu loves Jihae.' As in boyfriend loves girlfriend," and then I read off the other names of who loves whom.

Jungsan Valley was very much like Gollum. Plunge pools provided deep cover among submerged boulders the size of vw Beetles. Erosion—always present—seemed like an ancient event, and the slow-motion scouring created smooth, weird-shaped rocks and flow-form sculptures that dissipated the violent vertical-drop energy of the current. The water is fast, and there are magic seams where fast water meets slow water; it's the foam line, the bubble trail, the place to place your fly. The stream is fed by minuscule rivulets pouring out from the brush along the bank.

Stunted dwarf trees will grow if the boulders let them get a root-hold in the gravelly soil. There are few fallen trees in the water, no large woody debris, no weeds, no vegetation. It is pure granite. During bright, sunny days, success will be tough with dry flies, and it is better to work the rock overhangs of boulders embedded into the mountainside like smooth, polished warts. The only shade is from the valley itself and the looming boulders. The streams change temperature throughout the day, and that changes the behavior of the trout. The metabolism of a trout directly correlates to the temperature of the water. In winter, the trout feed very little and stay deep or swim near the sunlit surface yet ignore casted offerings. In some places the current flows underground through hidden passages, and you look between the crevices under your feet and see running water. It was here where a boulder slab shifted and pinned my boot and I sat for a long time, numb-footed, until I internalized enough focused strength to deadlift my extrication.

As I went farther upstream, there was less garbage, but more stone cairns appeared and these I left alone. I scented incense in the downwind breeze. I hiked around a bend and there, backed by bamboo and pine, was the severed head of a hog, and the incense wafted past its snout as if the pig were softly exhaling a bong hit. Spread out on the flat-rock slab was an opened bottle of soju, an apple, a pear, a husk of dried squid, and a lit candle.

Finding the chopped-off pig's head on a trout stream might not make any sense unless you have read David A. Mason's book *Spirit of the Mountains: Korea's San-shin and Traditions of Mountain-Worship*. It is a coffee table–sized book, heavy with photos and deep field research on the mythology of mountain shamanism. It is the kind of book where you know the author burned boot leather to get the goods. It is about a native religion quietly practiced at individual shrines and Buddhist temples across the country. San-shin is depicted as a white-bearded spirit-sage that inhabits the mountain valleys, creating sacred spots of spiritual energy. Mason digs into the totem symbols that surround San-shin: tigers, pine trees, and ginseng. Later I met Mason on a hiking tour he guided for the Royal Asiatic Society, and once he and his wife visited Jirisan, and I showed him some of the San-shin shrines I had come across. We kept in touch, and I would send him photos and field notes of San-shin shrines that I discovered in my explorations. Once you find a few, you see that they are everywhere.

Martin Russ described this in his war memoir, *The Last Parallel: A Marine's War Journal*, where he wrote: "Hills and mountains are looked upon as gods. I can understand this. The hills and mountains here are severe, grotesque, sometimes beautiful, always startling, each with a personality or characteristic of its own. In order to appease the hill and mountain gods, votive offerings are carried up to them."

Nearby was a small cave the size of a walk-in closet. It had been formed by a jumble of giant boulders, and the floor of the alcove was leveled by intricately placed flat rocks to form a low altar before which to prostrate oneself. Tucked in the crevices were numerous candles. Sometimes there was a person in the cave wearing a white robe and chanting and bowing. I never knew if it was a man or a woman, but I often saw old *ajummas* tending to the alms. I always skipped fishing that nearby pool to not disturb the pilgrim. The clients I guided on this stretch always found this fascinating, but one client was disturbed by it and said it reminded him of voodoo "cults" in New Orleans. He said he wanted to "get the hell away from these God-dammed witch doctors." It was always eerie to come across a candle that was still aflame or an incense stick that was still burning, knowing that someone else had been there not long ago or was still lingering around. It was like surviving alone on a deserted island and one day finding human footprints in the sand. When I started fishing Jungsan and Gollum Valleys, I found so many candles washed downstream by floods I filled two backpacks.

I emerged from a pool and reclined against a sloped rock and toweled off my hands. I wrote the trout numbers in my journal. The findings were about the same as Gollum Valley. Some pools held many trout; a few were scarce and others were a mix. I was pleased with these observations, and my anxiety over them being poached into unfishable numbers vanished. I stuffed away the journal and hand towel in my backpack, and when I looked down, a patch of sandy, fine-grain gravel had been disturbed. The gravel area was less than one square yard and surrounded by small boulders.

This was a trout stream for fugitives. If an outlaw were being chased by a posse, all he would have to do is to skip from rock to rock and never leave a track. There might be a time when he would be forced to wade through a pool or riffles and then clamber onto the next rock, but that would leave

Fig. 13. A San-shin pilgrim bows to the Mountain Spirit on a Jiri Mountain trout stream. Photo by James Card.

only a splashy trace that would quickly evaporate. There were few places to leave tracks, and they were pocketed patches and small shoals of sandy grit. The track was not human, and I looked more closely and measured it with my hand. It was about the same size. Above the wedge-shaped indentation of the pad of a hind foot, the toe prints were marked by small scratches in the grit formed by its claws.

I rockhopped back about twenty feet and studied the course of travel, imagining myself as a four-legged animal. The gravel patch was part of an open area that was easy to pass through. Tangles of boulders upstream and downstream made fording the creek difficult. I looked near the edge of the woods and looked over other patches of sand and gravel. There were no other tracks. The bears were the reason I had first come to Jirisan years ago as a hiker, with hopes of encountering one. Now cherry trout were the reason I came to Jirisan as an angler. Other than seeing the bears at Moon-su Temple, this paw print was the closest I had ever come to encountering one in the wild. I kicked around the idea of driving into Sancheong—the closest town that might have a school supply store—and buying some plaster of Paris and making a cast of the track as a souvenir. The track was too shallow and faint to get a good mold. The gravel-grit did not hold a deep indentation like the squish of loamy mud.

Most of the half-moon bears are thought to survive in Jirisan National Park. They have been hunted to near extinction for centuries. In traditional Asian medicine, the bear's gallbladders are considered to be powerful medicine for every malady known. There is a synthetic form, ursodeoxycholic acid, and gallbladders from domesticated farm animals can also be used, but that is not enough—the bile of wild bears is demanded. A bear gallbladder is worth more than its weight in gold and cocaine. Fraud is rampant, and Koreans swindle each other with similar-looking pig gallbladders.

I mulled over the purported efficacy of a natural substance so strong that people would kill a species into near extinction. What was strange was that they were spiritually killing themselves. The heritage of the bear is part of Dangun, the foundation myth of the Korean people. A heavenly father granted his son Hwanung permission to live among humans on Earth. A tiger and a bear asked him to transform them into humans. He agreed but they had to pass a test: to live together in a cave for a hundred days eating nothing but mugwort and twenty cloves of garlic. The tiger left early but the bear stayed, and Hwanung transformed the bear into a beautiful human female named Ungnyeo, which means "bear-woman" in old Sino-Korean script. Ungnyeo eventually desired a child, and she had no husband. Hwanung married her, and they had a son named Dangun, who became the ruler of Korea's first kingdom.

The bear is in their blood. I thought of the great bear clans of the Great Lakes Anishinaabe people. They were known as defenders and healers. The bear was their identity and was part of their "medicine," a spiritual healing presence embodied in a place, a person, an object, an animal, or an occurrence. There are medicine people that help others. There are medicine bags, pouches that hold sacred objects and personal totems. There are medicine wheels, ancient stone circles used for ceremonies with spokes pointing east, south, west, and north. Traditional Asian medicine took a different approach and found medicine within the bear itself and commoditized it.

These misty mountains, this rock-tumbled freestone creek, the bonewater maple trees, the parr-marked coppery trout, the bear track in the pebbly grit, all combined, were my medicine.

The decimation of the Asiatic black bear is South Korea's national shame, but now it extends internationally: having destroyed their own bear populations, they look to North America to feed their rapacious appetite. South Korea is the world's largest market for imported bear parts. Bear poaching rings have been busted in North America since the 1980s, and the reports always include the names of South Korean nationals and Korean Americans. They gut the bears, remove the gall, cut off their paws for soup, and leave behind the carcasses. The passage of time will eventually allow the bears to survive and repopulate. With rural depopulation and the mass inclination for urban living, few will be left who know how to set a snare or identify a bear track. After all, very few young South Koreans know how to even catch a fish. The old poachers' knowledge will die out. As internet-addicted urbanites, few young men will have the nerve to knife a trapped bear, slit open its guts, and remove its gall bladder. South Korean youth have the same problems as their American cohorts: they are delaying marriage, they are delaying buying a home, they are postponing or forgoing having children and starting families, they are in debt with student loans, they cannot find decent-paying jobs, and they do not have time to go looking for bears.

8 Field-Expedient Pattern Recognition

My client was a U.S. Army colonel based in Seoul at Yongsan. He was lean and fit and a good caster. He was a good soldier too, as he carefully listened to my directions on where to cast and how to approach each pool to get in place for the best casting position. He executed the instructions with aggressiveness and stealth. He steadily caught cherry trout at each pool and riffle. It was a perfect autumn day, and the sunlight was warm and the leaves were on the edge of turning colors.

The snake stood up, cobra-like, and flexed its tiger-striped back muscles. It was on a small patch of rocks that stuck out from the shoreline. The colonel didn't see it, and he was wading closer and closer to it—getting about eight feet away.

"Might want to watch out for that snake there," I said.

He looked over his shoulder, and his eyes popped as if he were about to vomit. He screeched and, raising his knees high, goat-stepped across the stream. Off balance the very first second, he lurched over the slippery rocks until he made it to the opposite shoreline.

I walked up behind the snake, paused for a moment, and snatched it by the tail. I flipped it out and away from me. It was about three feet long. I whipped the snake in an arc and rotated once, twice, like a cowpoke throwing a lasso, and launched the snake to the opposite shore, where it smacked against a boulder.

The client waded back across the stream, watching the spot where the snake had landed, twenty feet away. He was shaken up. His line was now tangled among rocks and woody debris in the stream.

"It's poisonous but it's harmless," I said. "It's a tiger keelback."

"What?"

"It's back-fanged. Say a snake like a rattlesnake or a cottonmouth or some other kind of viper. Their fangs are up front and as soon as they strike, those fangs are injecting venom into flesh. With the tiger keelback, there are smaller fangs in the back of its mouth. So, when it gets a hold of a frog or a rat, the venom is injected once the prey is deep in its mouth. You'd have to stick your thumb down its throat to get into trouble," I said.

"I fucking hate snakes," he said.

"You come across tiger keelbacks. I've taunted them with the tip of my fly rod and got them to rise up like that, like they are going to strike."

"It looked like a goddamn cobra!" he said.

"Looks tough, like a hognose snake back home, but it never strikes. I've teased these snakes many, many times and it has never attacked. All show and no go."

The first time I saw a tiger keelback flex like that, I was bass fishing some reservoirs around Jinju. The snake was right behind me. I holy-shitted myself just like the client.

The keelback was on top of a rock on the opposite shore.

"Look, there he is," I said. The snake slithered off the rock and into the water. "It's coming back."

"Son of a bitch," said the client as he backed up the bank.

I grabbed him by the waders. "Hang on. Just watch," I said.

After the snake swam about halfway, the main current pulled it down into some riffles, and those riffles pulled it into some faster water, and then it tumbled into even faster water and disappeared.

I learned these things from Stephen Karsen. I wrote an outdoor column for the *Korea Herald*, and he contacted me as a fellow American expat interested in Korean flora and fauna. He asked if I wanted to go "herping." His passion was finding reptiles and amphibians. My passion was finding freshwater fish. I agreed to meet him and figured I might come across some new creeks during our rambles.

Karsen was from Illinois and lived with his family in Daejeon. He was an environmental science teacher and dorm counselor at Daejeon Christian International School. His students received hands-on education during field trips, and he brought captured (and later released) frogs, turtles, and snakes into the classroom and sometimes some road-killed critters to dissect.

I met him at his house, and we drove into the countryside to a dull-looking village nestled into an equally dull-looking valley under Jangtae Mountain. I was used to scrambling up waterfalls on pristine mountain streams overshadowed by jagged cliffs, but this was good upland country, just plain and remote enough to get away from the city. We talked about every species of creature on the peninsula and queried each other on our sources and field observations. We parked by the village and headed into the valley.

Near a farmhouse there was a weedy pile of scrap wood and corrugated tin—the kind of materials one probably uses once for a makeshift shed. There was a burrow, some scratched earth under the tangle of boards, and metal. "There's got to be something in there," said Karsen and knelt in front of it. He wore a T-shirt and no gloves. He stuck his hand and arm into the hole, all the way to his armpit. My stomach turned a little. He grunted with happiness and lunged back, gripping a large, writhing snake.

He said it was a Dione's rat snake, and it was not poisonous. It was about thirty inches long, and its camouflage was perfect: a mix of green and brown with black bars. He said he found them often and this was a good specimen. He admired it for a moment and released it back into its hole. I immediately liked Karsen. It was the way he intuitively knew with great confidence that there would be a snake under that pile of junk. I liked that he had the balls to reach bare-handed into a hole that he could not see into and pull out a scaly reptile with joy, not fear.

We hiked onward to a small creek. The rivulet was too small to hold any game fish, but it was clean, cold water and the banks were lush with foliage. He found a Korean fire-bellied toad (*Bombina orientalis*) and flipped it over. Its belly was bright orange and mottled with black spots. He said the slime on them is toxic and to not touch my eyes. Karsen tipped over rocks along the creek and found an Amur grass lizard, a small skink-like creature small enough to hold in the palm of his hand and with a tail nearly twice as long as its body.

He lifted another stone and found a snake that made him pause. I knew what it was from the wedge-shaped head and thick-angled spine. It was a pit viper. The Koreans call it *salmosa,* the snake of four steps, as in it strikes, and you take four steps and fall over dead. Brown spots ringed in black covered its rusty tan body. It was about a foot long. "He won't bother you. Leave him alone, he leaves you alone," said Karsen. The viper slithered away and vanished under some rocks. I walked over to where it had disappeared and snapped a twig to mark the spot as a place to avoid.

Karsen continued to turn over rocks near the creek. He said he wanted to show me something. "Ah, here it is," he said. "No, not this one. But take a look anyway. This is *Leechii hynobius,* the most common salamander in Korea."

It fit in the palm of his hand, and he handed it to me. It had black speckles over its olive-green body. I held it and looked it over while Karsen kept turning over rocks.

"Here it is! Come here. Look," he said. He was giddy. I tucked the salamander under a damp stone and walked over to Karsen. He was holding another salamander. It was about 1.5 inches long, with a dark body and a red-orange stripe on its back. On its underside were tiny white speckles.

He said it was unknown to science. I asked him what he meant. It was a lungless salamander that breathes through its moist skin. It is terrestrial, and it lays its eggs on land. All other Asian salamanders breed in water. He explained that this salamander is not in guidebooks nor does it have any scientific descriptions. Karsen said he contacted his old professor at Southern Illinois University on how to proceed.

The next time I visited Karsen, he was with Dr. David Vieites from the University of California-Berkeley. He was there for a two-week trip to collect specimens and tissue samples for DNA analysis. "This is the most important event in the history of modern herpetology," said Vieites.

My job was to help turn over rocks and find what would be later called the "Korean crevice salamander." What makes this salamander special is that it is not where it is supposed to be. It would be like discovering a hidden population of native snow baboons in the northwoods of Wisconsin. It was a zoological enigma. This species of lungless salamander was only known to exist in North and South America. There was one small population in Italy and Sardinia, and that odd group left biologists pondering how they got there. Now that Karsen had discovered another population in Korea, this

Fig. 14. American science teacher Steven Karsen discovered the Korean crevice salamander, *Karsenia koreana*, and it was considered the biggest event in modern herpetology. Photo by James Card.

makes the biogeographical and evolutionary possibilities more intriguing. Vieites said he was already making plans to look for it in China.

Dr. David Wake, one of the world's foremost amphibian experts, formally introduced the salamander to the world in the May 2005 issue of the prestigious science journal *Nature*. Wake called it "the most stunning discovery in the field of herpetology during my lifetime." The salamander is not only a new species, but also a new genus of the family *Plethodontidae*. The creature was named *Karsenia koreana*.

Ever since those field trips, I thought of Steve Karsen and *Karsenia koreana* every time I tried to match the hatch while fly fishing. I had a guidebook to Korean insects, and there was a section on mayflies, caddisflies, and stoneflies. Over time I was able to find and identify each species, but there were times when I came across specimens that did not match up with any of those listed in the guidebook. Like the salamander, they simply had no written record of existence. The differences between the known species and newly observed oddball species were subtle: some caddisflies were too small to match up with the ones in the guidebook: one mayfly had weird coloration, and another stonefly was unnaturally huge, and so on.

I cultivated daydreams of having an aquatic insect named after me like Karsen's salamander. Not only that but I would develop a fly pattern for it as well. But unlike Karsen, who could wander the creeks and flip over rocks and find his namesake anytime he wanted, I was playing a different game. Karsen could find his salamander easily during the warm-weather months, and he could probably find some hibernating in the winter too. For me there was a very limited window of opportunity to recognize, identify, and capture an unusual species of mayfly, caddisfly, or stonefly as their hatches appeared only at brief times of the year.

There was no hatch chart, so I had to make one. There was the very common pattern mayfly (*Ephemera strigata*), the sunbeam (*Heptageniakihada*), the old-days mayfly (*Siphlonurus chankae*), the fan and the spotted fan (*Epeorus pellucids* and *Epeorus latifolium*), the spring virgin (*Cinygmulagrandifolia*), the flathead (*Ecdyonurus levis*), the pinstripe (*Ephemera separigata*), and the river mayfly (*Rhoenanthus coreanus*).

For stoneflies, there was the native-named Korean stonefly (*Kamimuria coreana*), the true stonefly (*Oyamia coreana*), the big-water (*Pteronarcys macrasachalina*), and the pygmy (*Rhopalopsole mahunkai*). For caddis, there was the big stripe (*Macronema radiatum*), the small sedge (*Nemotaulius brevilinea*), and the whiskered (*Stenopsyche griseipennis*).

My hatch chart never turned out to be more than a half-filled grid with a jumble of field notes. Sometimes the bugs emerged, and sometimes they did not and my hatch chart was dotted with incomplete observations. I kept coming across insects that were not in my guidebook. Sometimes I snapped a close-up photo if one specimen paused on a boulder. After a few seasons I was able to recognize the usual suspects, but more often than not, I came across a species that I had never seen before. I bought a butterfly net with the intention of collecting and preserving specimens for better identification. I used it once and thereafter it stayed in the car. It was one more thing to carry to the trout stream, and I followed the dictum, "The hunter who chases two rabbits catches neither one," and I always decided to chase the trout.

Fig. 15. The strigata mayfly, *Ephemera strigata*, is one of the most common mayflies on cold-water creeks. Photo by James Card.

I came to the conclusion that most known and invented fly patterns will work just about anywhere on Earth. Some will naturally work better during certain times and places, especially if a hatch is occurring. I tossed the same patterns that any fly angler would have on a North American stream. The trick was to keep a fly box loaded with a variety of mayfly, stonefly, and caddisfly patterns that could be used to imitate some of the native aquatic insects in size and color. Some chunky orange simulators mimicked one stonefly quite well. There was a mayfly that resembled a blue-winged olive, and they were hatching on a lenok stream one day in October. That was all they were feeding on, and it was one of the best days for lenok I ever experienced. I never encountered that insect again. It was a meeting of pure chance, and I happened to have that pattern. Some unlucky patterns were relocated farther and farther to the edges of the fly box.

My go-to patterns were always caddisflies. They never let me down. My most useful tool was a black permanent marker. Black caddis were everywhere and very common throughout the peninsula, so I bulk ordered elk-hair caddis of various sizes from a Chaing Mai, Thailand, fly maker and then blackened them with the marker. One of my favorite tricks was to use a coffin fly to mimic the common *strigata* mayfly. The coffin fly was an imitation of the green drake spinner, and the thorax and abdomen were white. The *strigata* was similar but had black bars, so to mimic it I dotted it with the marker. All pale-colored mayfly patterns were useful as I could always smudge them to match the hatch with some colored markers.

I had a fly-tying kit, and I tried very hard to get interested in this related craft. Many of my days were spent writing: sitting at a desk and applying my concentration and attention to detail for hours on end. After a stint of that, I wanted to get outside and go fishing or do something physical, not sit at a desk tying flies and burning through more of my limited reserve of concentration, which I needed to protect for the next day's work. I learned the basics and that was enough. My only fly-tying accomplishment was making some caddisflies with hair I pinched from a road-killed water deer.

9   Under the DMZ

I woke up an hour early because of the screaming. I could not fall back asleep, so I crawled out of the tent and fired up the backpacker stove to make some cowboy coffee under the beam of my headlamp. There were two sounds: an aggressive snarl and a desperate screech. A predator was killing its prey. I was camped on Phantom Creek, and I had nicknamed it that because I'd interviewed an old farmer who was watering his Hanwoo bull in the creek. He was a gentle chestnut-colored beast, and he followed the farmer around like a dog. I asked about the existence of trout. He confirmed there were lenok, but he also said the valley was haunted. He said the word *gwisin* (ghost) many times while pointing upstream. He said that the deeper one goes into the valley, the better the fishing, but the more haunted it becomes. I figured the fishing up there must be good to make up a story like that.

The afternoon before, I had left a cage fight in Seoul. It was South Korea's very first mixed martial arts match, held in an octagon ring made of chain-link fencing and held at Changchung Gymnasium. There was a purse of 50 million won. Any fighting style was welcome, and there were very few rules. Lee Myeon-ju, a Muay Thai kickboxer, went three extra rounds with Lee Eun-soo to win the championship match. Blood dripped out of both fighters' noses, and the mat was smeared red. In one of the rounds, Eun-soo straddled Myeon-ju, pinned him with his body weight, and rained down punches on his head.

Blood mist sprayed across the canvas, and a speck of blood landed on my reporter's notebook. The punches made a sickening sound of bone smacking hard against human flesh. The screams in this haunted valley reminded me of this sound, of one animal destroying another. After the fight, I went to a *dabang* coffee shop, wrote up my notes, filed my story for the *Korea Herald*, and left for the mountains.

The screeching continued. It was the sound of a deranged animal but also slightly human. It was not a canine sound. It was feline. I threw on a persimmon-dyed work jacket that my wife had given me as a gift. The cotton was soft, and it buffered the predawn chill. The hour before sunrise is always the coldest. I unsheathed my cruiser axe and slammed the bit into a dead log, leaving the haft sticking up for a quick retrieve. I tended to my pot of coffee and set my spotlight near the axe. This is what soldiers guarding the DMZ got to listen to during night watch, as if the specter of infiltrating North Korean commandos weren't enough to put a man on edge.

As a freelance project, I copyedited the English translation of Hahm Kwang-bok's book, *The Living History of the DMZ: 30 Years of Journeys in the Borderlands*. He had spent much of his life exploring the area. I learned there are abandoned train stations, churches, roads, rail lines, schools, mines, and bridges left behind in this stretch of no-man's land. The only people who got to see the old architecture and infrastructure were soldiers who happened to be assigned to patrol that area. Perhaps that is what the farmer was referring to, that farther up the valley was a haunted house. It was unlikely unless the house was carved into the mountain. The valley was tight, almost ravine-like, and the creek was skinny and shallow. It was a tributary of another tributary that flowed into a river that held lenok. My theory was that if they were in that water, they would be in these waters. However, while blue-lining the tributaries, I found numerous weirs and dams that made fish migration difficult. I was looking for a population of fish left alone and isolated and forsaken, much like the abandoned villages in the DMZ. A relic population of fish cut off from the rest of the world during the postwar years, when South Korea industrialized and set about damming rivers and building roads through the mountains. These were the trout that the world forgot.

I was as far north as one could legally get in South Korea, without getting into trouble with the law or the militaries of two countries engaged in a never-ending war. The thirty-eighth parallel was established by the U.S. War Department in the last days of World War II. With the Japanese no longer occupying the peninsula, it was a geographic power vacuum, and the thirty-eighth was a line in the sand to thwart Russian advancement. In 1950 Communist-backed North Korea attacked South Korea, creating the Korean War, and three years later, the thirty-eighth parallel still remains as the world's longest and most dangerous line of scrimmage.

In Wisconsin and other parts of the Midwest, people discuss the concept of heading "up north," and everyone has a different definition of it. In the mind's eye, it is a place of endless forests, pristine lakes, rolling rivers, and cozy cabins. One thing is certain about being up north: it is a place where the fishing is always better, and that was why I was on the edge of the DMZ. It was as far north as I could go.

In Wisconsin, a tension zone stretches across the state, merging the Northern Mixed forest of what we call the northwoods with the Southern Broadleaf forest, a mix of woods that was once forest, oak savanna, and prairie and is now what we call farm country. As I drove north to the DMZ from my home on the southern coast, I saw the same transition. I lived in the Southern Evergreen forests, where a mix of pine, laurel, and bamboo grew along the coast, but that quickly transitioned to the Central Korean deciduous forests, which are composed of hornbeam, maples, and oaks and areas that were repopulated with pine, cedar, and acacia during South Korea's reforestation drives in the 1970s.

About halfway up the peninsula and somewhere past Andong, you hit a tension zone where the forests of the south meet with the Manchurian Mixed forests. The names of the trees that populate these woods reflect their range: Korean pine; Amur linden; Manchurian ash, fir, and elm; Mongolian oak; and Siberian spruce.

At the northern edge of the Gangwon Province, near the DMZ, I was at the gateway to Asia's northwoods, and from here to the Arctic Circle it held some of the greatest freshwater game fish on Earth. It is the Ultima Thule of the East, the hyperborea of Asia. There was the black-spotted pike *Esox reichertii* (the American northern pike has white spots). There was the king of all salmonids, the Siberian taimen, a monstrous trout of dreams and nightmares. That fish was part of the *Hucho* genus, and similar salmonids were scattered throughout northern Eurasia. There was the East

Siberian grayling (*Thymallus pallasii*), one of many species of grayling in the region. There were sturgeon, shad, smelt, whitefish, and salmon—many of those species were the same ones found in the rivers of America's West Coast. But there were native ones, like my beloved cherry trout, with its own family of anadromous and landlocked cousins.

Most interesting were the true piscine symbols of the northwoods: the *Salvelinus* genus, also known as the char family. The most famous member of this group is America's brook trout, considered to be a jewel of any cold-water creek. Northern Asia has many in this family, including the *Salvelinus leucomaenis*, the white-spotted char, found in Japan and Russia. There is a subspecies called the kirikuchi char (*Salvelinus leucomaenis japonicus*) that lives in the Kii Peninsula of Japan. It is the southernmost population of any *Salvelinus* and is considered a remnant population that has survived in that isolated watershed since the last Ice Age or the Pangaea. In my research, I found a few references to *S. leucomaenis* being recorded on the Korean Peninsula, but I concluded any that might exist would be in North Korea. The existence of them in South Korea would be an undiscovered longshot.

I was at the threshold of a great northern ecosystem that held incredible adventure and fish that were exotic yet familiar, and I was blocked by the world's biggest stalemate. The DMZ is the world's most heavily fortified border. It stretches across the peninsula for 151 miles, from the east coast to the west coast. It has 1.2 million landmines on the South Korean side and thousands of soldiers facing off across mountain valleys. The DMZ is composed of multiple parallel lines: some are invisible boundaries, and others are chain-link fences crowned with concertina wire. Some of the fences are energized to electrocute infiltrators. Pinched into the chain-link fencing are small rocks. If a rock is disturbed, it is a sign of a potential intruder. Agent Orange was heavily used here in the 1960s. Both the South and North Korean troops conduct controlled burns to keep the brush down and to better their sight lines, and their fires set off landmines. The last U.S. soldiers killed while serving on the DMZ died in 1976, over a tree-pruning dispute. They were axed to death by North Korean soldiers.

Besides trout, my freelance journalism brought me to the DMZ. While on assignment for *Earth Island Journal*, I got access to the Dragon Moors, a rare alpine wetland designated as a Ramsar site. I got permission to visit in winter after a fresh dusting of crunchy corn snow. Accompanied by soldiers, we rode in a jeep and passed through the Civilian Control Zone, past the southern defense lines, and past barbed wire fences adorned with landmine warning signs. We went up a forested valley with switchbacks until we reached a ridge 3,900 feet above sea level on Daeam Mountain. We hiked down into a depression, where this high-altitude peat bog evolved among the mountain peaks over 5,000 years ago. It was there where I truly got a good look at North Korea, the kingdom of the autarckic Kim family, who ruled a cold and harsh country with the iron fists of a hereditary Confucian-Communist mafia.

Their mountains were in the distance, and I expected to see jeeps or tanks or tiny figures of soldiers through my binoculars. Instead, there was nothing, just pure quietude. Looking at the horizon I imagined MiG-15s and F-86 Sabres hunting each other on the distant skyline in the early days of jet-to-jet dogfights, and of Ted Williams flying his F9F Panther on bombing missions behind enemy lines. It was the most heavily armed strip of land in the world, and I found it to be the most peaceful place on Earth. I looked at the mountains of North Korea for a long time, and they stretched as far as I could see, off into the horizon and into Asia's north country. On nighttime satellite photographs taken of the peninsula, South Korea is bedazzled with light, and North Korea is blanketed in darkness with only the capital, Pyongyang, having a mote of a glow. The DMZ is the dividing line between light and darkness, freedom and totalitarianism.

For the *Wisconsin State Journal*, I interviewed soldiers from the Badger State serving on the DMZ. I met them at Camp Bonifas, home of the "World's Most Dangerous Golf Course." It is a par three course with one hole, it is 192 yards long, and a sign warns, "Danger! Do not retrieve balls from the rough—live mine fields." A private from Oshkosh told me soldiers flushed water deer on patrol, and the sounds of exploding landmines were usually attributed to animals, but one was never certain.

This was the kind of country you could hike across for miles without seeing another human, but when you did, it would be a South Korean soldier with a Daewoo K2 leveled at you or a North Korean soldier with a Type 8 (an AK-47 derivative). This was the country where my Uncle Clayton served during the Korean War. I kept looking for the bridges where Red Jensen welded his name. I did not find them and this was disappointing, but finding trout made me feel better. The only bridge-like remnants from the

Korean War that I found were Marston mats—perforated steel planking—and these strips had been repurposed as foot bridges over creek crossings.

I sipped on my coffee and kept an eye on the horizon above the tree line. Soon the ooze of nautical twilight would emerge and I could turn off my headlamp, and then the orange-pink dawn would filter through the tree tops as the sun rose over the East Sea. The screaming and screeching stopped, and it was quiet. I removed the *jangdo* from my fly vest to slice up a *chamoe* for breakfast. It was an odd little fruit that was like a honeydew melon with a taste of cucumber. The knife was given to me by master bladesmith Pak Yong-kee in Gwangyang. He was the only person left who knew how to make this traditional knife. During the Japanese occupation, carrying a knife was outlawed, and Pak mastered the forbidden art of knife making. He traveled across the country to find *jangdos* in museums so he could study their construction. The slender gentleman's blade was three inches long, and Pak forged it on an anvil and honed the edge with a file and whetstone. Etched on the blade were stylized lines meaning "dragon," the meaning of his middle name, Yong. This was his artist signature.

I passed the time until daylight eating the melon and playing the process of elimination. I was certain it was a big cat. Only three animals could make such a sound: a tiger, a lynx, and a leopard. Finding the existence of a Siberian, or Amur, tiger (*Panthera tigris altaica*) in modern-day South Korea would be like discovering an ivory-billed woodpecker, the Lord God bird, which is thought to be extinct in North America, but rumors of its existence still persist. In the old days of preindustrial Korea, the tiger ranged throughout the peninsula. Tigers were to be feared back then. In the early 1800s tigers killed several hundred people. The decline of the tiger began in the nineteenth century as it was heavily hunted for its pelt, which was traded and exported. Habitat destruction took its obvious toll, along with the introduction of the rifle. The last tiger killed in Korea occurred in the Daeduk Mountain area, near Muju, in 1921. It was killed by a Japanese hunter. The last confirmed tiger sighting in South Korea was in 1946 in the Sorak Mountain range.

It could have been a Eurasian lynx (*Lynx lynx*), but historically they have ranged in what is now northern North Korea and farther north. It is very much like the Canada lynx, so I figured its range would be similar: that lynx is found in very few places in the lower forty-eight states. Like its Canadian cousin, it has long legs and big paws that act like snowshoes, for traveling through deep snow.

Two other predators, the Manchurian grizzly (*Ursus arctos lasiotus*) and the gray wolf (*Canis lupus chanco*) could be eliminated from the list. The Manchurian brown bear is thought to live on the northern borders of North Korea, and perhaps in the Baekdu Mountain range, but it is currently considered endangered or nearing extinction because of being hunted down for its gall bladder. The extermination of gray wolves in Korea corresponds almost exactly to the decline of the wolf in North America. At the turn of the century, Koreans hunted and trapped them just like their American counterparts. Just as the wolf was widespread in North America, the gray wolf in Korea once ranged throughout the peninsula. By the 1960s it was extirpated. Like the brown bear, if any populations exist, it is mostly in the Baekdu Mountain region.

That left the leopard. The Amur leopard (*Panthera pardus orientalis*) is thought to be extinct in South Korea, and their current range is the forested terrain of Baekdu Mountain, which straddles the border between China and North Korea; and the Tumen River, an ecological corridor of the Russian, Chinese, and North Korean borderlands. It is one of the most endangered feline species in the world. The last leopard to be captured in South Korea was in 1969 on Odo Mountain, in Hapcheon County in the Gyeongnam Province.

P.p. orientalis is adapted to the northern extremes. It has the longest legs of the leopard family, which helps it walk through deep snow. Its cream-colored winter coat grows long, and the Amur subspecies tends to have larger rosettes than other leopards, with unbroken rings and dark centers. The rosettes allow them to blend into their woodland surroundings as they hunt. A nocturnal hunter, the Amur leopard stalks its prey until it is close enough to burst upon it and incapacitate it with a killing bite to the neck, while using its curved claws to bring it down. They are also ambush hunters from the trees and rock ledges and will wait for unsuspecting game to pass within range.

A handful of Korean ecologists and wildlife aficionados believe the tiger and leopard might still exist, and if it does it is most likely in the DMZ, but wild animals do not recognize geopolitical boundaries, only good habitat. The DMZ is the world's largest accidental nature preserve.

While I was living in Jinhae on the naval base, there was a rear pedestrian entrance guarded by a young soldier. He would always salute me as I slipped out the back, even though I was just a civilian English teacher. Before me were the foothills of the great Amin ridgeline, which separated the cities of Jinhae and Changwon. There was one winding road that went over the top, but most of the traffic passed through a tunnel bored through the mountain. Other than that, it was a massive roadless area with many hiking trails, and I found the area to have many water deer.

I hunted them with a camera with the hope of getting some publishable photos. This was when I found the tree scratching. Big cats use trees as signposts to signal their territory and when a female is coming into heat. Urine spraying and defecation are also used as marking communicators. The hooked, retractable claws collect bacteria and debris, and by scratching a tree, the claws are cleaned and honed.

There were two of them on small broadleaf hardwood trees that I never identified. The highest point of the scratching on the tree reached five feet off the ground; the other was four feet, seven inches. The thin, hard bark was shredded, and the sapwood was exposed. I studied it for a long time and took a photo. There was no other explanation. It was away from the main hiking trails, and there was no human activity in this forested part of the ridge. If a human were to do this, the only tool I could imagine being used is a handsaw, tilted at an angle and violently raked downward, but in a way that the blade did not cut into the wood. But why would anyone do such a thing?

In my scrapbook, I saved an article by Han Sang-hoon, a wildlife researcher at the Ministry of the Environment. He published an article in the magazine *Man and Mountain*, where he posited that there were thirty leopards on the Korean Peninsula and at least ten were in South Korea. His conclusions were based on eyewitness observations, scat, and other signs. The scratched-up trees I found on Amin ridge were a sign.

I broke camp at daylight. As I hiked into the narrow valley, I thought of what the farmer said about the valley being haunted. I thought of the film *The Ghost and the Darkness*, based on the book *The Man-Eaters of Tsavo* by Lieutenant Colonel John Henry Patterson. Then I thought about the month I spent in India and read the *Man-Eaters of Kumaon* by Jim Corbett, which had me on edge while fishing for mahseer.

My old friend Robert Neff, a historian who studies the late Joseon period (1880–1930), wrote a series of articles about man-eaters in his regular *Korea Times* column. He found that in the mining districts of North Korea, gray wolves killed more people than tigers and leopards. They killed twenty-eight people in 1928, many of them children. The leopards also killed many people, sometimes hunting them from their rooftops.

Neff discovered an old Chinese saying, "The Korean hunts the tiger six months of the year and the tiger hunts the Korean the other six months." The tigers would walk into villages and snatch victims, sometimes going right into homes. Korean tiger hunters were known for their bravery and woodsmanship and were their own paramilitary force, wearing blue uniforms and tiger and leopard skins. There are some early accounts of Westerners hunting on the Korean Peninsula, including Kermit Roosevelt, son of President Theodore Roosevelt. The younger Roosevelt hunted Korean tigers during an Asian expedition and referred to them as the "great invisible."

If tigers and leopards still exist, their travel corridor is the Baekdu-daegan, the unbroken chain of peaks, valleys, and ridges that stretches from Mt. Baekdu on the North Korean–Chinese border to the Jiri Mountains in the south. Within this mountainous "backbone" are subranges of smaller mountains, and the headwaters of all the rivers emanate from the watersheds within.

Adding to my unease was that I was not sure if I was within the DMZ, breaking the law. On the road that I drove in on, there was a weathered sign that read: "No Entrance" and stated that if one wants to visit the area, one must receive official permission. Entering without permission could lead to a 200,000-won fine. That dirt road meandered into the *mintongseon*, or the Civilian Control Zone of the DMZ. The trout stream I was interested in scouting was farther down the road, but it meandered in the same direction. None of my maps were of help because South Korea censored all military landmarks.

On a road map produced by the Ministry of Construction and Transportation, it appears that one could hop in a car and cruise up Highway 1 to Gaeseong in North Korea and visit the joint North-South factory complex. Similar maps were distributed by the Korea National Tourism Organization during the hosting of the World Cup. On these maps, the DMZ doesn't exist, and Korea is not a divided country.

A semi-honest map is produced by the Yanggu County government. The DMZ is marked in bold red letters on a perforated line. There is a parallel shaded area to represent a restricted space. I once visited Dutayeon, a scenic waterfall and plunge pool within the Civilian Control Zone. I had to be accompanied by an official tour guide after advance reservation, signing documents, and producing identification. It is a beautiful trout stream that is strung with barbed wire tagged with triangular landmine signs. At the entrance checkpoint, all cameras must be volunteered to the soldiers for security and safekeeping.

Along the Seohwa River, which flows out of the DMZ, army camps outnumber the farms. The tourist maps of the two counties that abut the DMZ, Hwacheon and Cheorwon, are especially deceitful in their misrepresentation of the DMZ boundary lines. Created by Beetle Maps, they portray a land where the DMZ zones do not exist, and it all looks like open country. The maps tend to flaunt their proximity to historical sites within the DMZ, yet those sites are inaccessible.

I walked by many patches of pocket water. The strip of water was narrow, and it would not sustain trout life, even for the smallest of fish. This was a lesson I learned early on while blue-lining creeks. When exploring new country for trout fishing, it is easy and obvious to stop at a bridge and look upstream and downstream. If the water does not look promising, you move along and cross that stream off the list. This is a mistake. The view from the bridge is a snapshot of the river and nothing more. The character of rivers can change very quickly, and dead water in one spot can change into perfect habitat in one hike around the bend.

The small riffles of pocket water led to a small pool formed by a knee-high waterfall. There was no need to rush. The trout were there. How you go about catching a fish is more important than bringing the fish to the net. I watched the water for a few minutes. I slipped to the side, crouched, and flicked a quick cast near the bubbles. The deer-hair caddisfly hung in the film for a second, and a silver-bellied fish snatched it. I walked over to the pool and netted a small lenok that was only five inches long. I held it in my hands and looked over its olive brown body speckled with dark spots. Its coloration wasn't as flashy or pretty as other trout, but it was perfect camouflage, adapted for surviving the cold-water streams of the Asian northwoods.

Fig. 16. A river flowing from inside the DMZ and protected by landmines and razor wire. Photo by James Card.

I slipped the little fish back into the water and noted that this fish had become my new mistress. For a long time, the cherry trout in the Jiri Mountains were like that: a passionate love whom I wanted to spend every second with. They were still very special to my heart, but the lenok were pulling my attention away. I was irrationally driving nine hours north to the DMZ every other weekend to hold these fish in my hands, and the fact that they lived in rivers that flowed out of a no-man's land militarized zone made them that much more of an object of desire.

I came to think of lenok as the brown trout of northern Asia and the Russian Far East. The brown trout is the founding fish that formed the sport of fly fishing in England, and I approached lenok as I would the venerable brown trout. The cherry trout, however, is in the *Oncorhynchus* genus in the family of *Salmonidae*, and the most famous and widely stocked of those species is the rainbow trout. I came to think of Korea's landlocked cherry trout populations as being like the cutthroat trout or the Little Kern golden trout, or the Gila trout or the Apache trout. They were cut off from their migrations to the sea by ancient geographic forces or, more recently, by the dam building of men. The commonality was they were lost souls living in remote and rugged rivers and only to be observed by those willing to put in the work to find them.

Lenok are fish of the *Brachymystax* genus, and they are often called Manchurian trout. There are only four of them. The most widespread are the sharp-snouted lenok (*B. lenok*), which ranges from Russia, Mongolia, and northeastern China. There is *B. savinovi*, which is endemic to eastern Kazakhstan. The species of my pursuit was *B. tumensis*, also known as the blunt-snouted lenok. The last name refers to the Tumen River, which forms a border between China, Russia, and North Korea. The sharp-snouted subspecies is the lenok of the northern range; the blunt-snouted is the lenok of its southern range. These areas overlap and hybrids have been recorded. There is also *B. tsinlingensis*, a Chinese subspecies from the Qinling Mountains, which form the Yellow and Yangtze River watersheds.

The names "sharp" and "blunt-nosed" are because the fish has an overbite, which is less pronounced in the blunt-nosed species. It is not an anadromous fish. It was landlocked during the last glacial age and has lived in cold-water inland rivers and streams ever since, creating their own primordial branch of the trout family tree. They do not have any migratory characteristics other than heading upstream to spawn after the spring ice melt and sheltering in deep holes when stressed by drought or warming temperatures.

The Korean name for the fish, *yeolmokeo*, means "fish with heated eyes." I sometimes called it the leper fish. There are a couple of legends about this fish. There was once a village of lepers that lived along Moondeung Stream, which flows from the Dutayeon waterfall, within the DMZ. North Korean soldiers—who wanted the village leveled—transformed their village into a fake military camp, and American bombers blew the place up. The lepers ran into the river to evade the firebombing and drowned. They were reincarnated into lenok—with reddened eyes from their tears.

In Hahm Kwang-bok's book, he tells a story of soldiers calling lenok the "Kim Il-sung fish," after the North Korean dictator who was the father of Kim Jong-il and the grandfather of Kim Jong-un. The DMZ has always been rugged, porous country infiltrated by enemy agents, and the soldiers

Fig. 17. South Korea has the world's southernmost population of Manchurian trout, or the blunt-snouted lenok, *Brachymystax tumensis*. Photo by James Card.

joked that the fish's eyes were bloodshot from lack of sleep as it swam all night downstream to spy on them.

They can grow up to around twenty-seven inches or even more in the right conditions. The largest I have caught have been in the ten-to-twelve-inch range, on small creeks, with a few in the teens that came from dark-water holes on bigger streams. Cold, shady, and highly oxygenated water is the key to their survival.

At the head of the pool was a ghost pine. Maybe this was a clue to why the farmer said the valley was filled with haunted spirits. It was a pine tree that didn't belong here. It was a lace-bark pine (*Pinus bungeana*), and the tree had been brought into Korea from China hundreds of years ago. I saw a few in Seoul, where they were designated as natural monuments. They are easy to identify as the bark exfoliates, leaving pale patches on the trunk. The Koreans call it the *baikkol-song*: the white-bone pine. I hiked past the pine and found more pocket water and small riffles. I cast into them and caught another small lenok. I continued upstream, and the creek became thinner and the water shallower. Far up on the mountain slope, fabric was snared on the branches of a pine tree. I looked at it for a long time trying to figure out what it was. I had no intention of scrambling uphill for three hundred yards to find out. I guessed it to be a North Korean propaganda balloon. There was no other explanation. Judging from the position of the sun, I was also deeper into the Civilian Control Zone of the DMZ. There was nothing haunting about the steep valley, as the farmer claimed, but the unease of being in a place I wasn't supposed to be kept me looking over my shoulder.

Although this small creek held lenok, and I could stab a green pin on my wall map marking more discovered water, it was an empty achievement. My solivagant angling expeditions to this area were leading to burnout. Mostly it was about logistics. For the past couple of years, I had been driving nine hours to get up into DMZ country. I would fish for as long as I could until I had to drive back for work on Monday morning.

I fished Buk Creek, a steep and narrow stream loaded with cherry trout that flows from Jinbu Mountain into the East Sea. I caught cherry trout in the Yeongok Stream, near Odaesan National Park, and lenok in the Odae Stream, which flows from the park. The creek was very small and fast and dropped at a steep gradient. The long-stretching valley was exceptionally scenic and, as a result, it was overrun with guesthouses that do not follow set-back rules because there are none. Plots of barren ground indicated more construction was to come. I crossed this stream off my list.

I loved the Seohwa River, north of Inje, near Pyeonchon Village, where I caught lenok and rainbows. It was bigger and burlier than the smaller streams, and it was on this stream where I caught my first Far Eastern catfish on a woolly bugger. It was like a long, skinny bullhead. Army camps were scattered throughout this area, and soldiers in jeeps were always on the road. I fished the clear and cold lenok water of Simjeokri Valley where Inbuk Creek and the Seohwa River converge.

I fished Suip Creek near Hyeonli Village. Its upper headwaters flow out of the DMZ, but once it passes the Civilian Control Zone it is open season. It had beautiful rapids and riffles, and I never caught a trout, only smashmouth perch. They were fun to catch in a small creek setting, but it left me disgusted to drive so far to catch a species of fish that I could have caught a short drive from my southern home.

I caught lenok and cherry trout in Naerin River, in the Piasi Valley near Saengdun. I came across sogari fisherman casting around and more weekend sightseers. I learned of two Korean suffixes that helped me make sense of the landscape: *dun* is the arable flatland at the base of a mountain slope, and *gari* is flatland in a valley where people can live and garden. Land that is not sloped is precious in this country. The Bangtae was an upper branch of the Naerin, near Kirin Village. It ran through a forested valley, and leonk were in there.

The Jawun and the Gyebang Streams in Hongcheon County were a bust—too many guesthouses crowded into the area. The mountain hostelry business was ever-expanding into the mountains as rural folk tried to find viable incomes. Renting out rooms to tourists was the easiest cash. The problem was that everyone had tried this, and some mountain valleys had more guesthouses than guests. It was a slow-motion development race to the bottom. I caught lanky lenok in the Jawun, along with escaped rainbows, and left after a couple of hours of fishing. I caught lenok in the Huengjeong Stream, near the buckwheat fields of Byongpyeong, and the best part of that trip was watching the town's cockfighting tournament with native roosters.

The Kiwha, a branch of the Dong River, is a rainbow river due to the presence of many trout farms raising rainbow trout that harness the abundant spring water. The Kiwha is a cement-lined river with a road next to it throughout its course. There are many signs that say No Parking, No Camping, and No Cooking, and a couple of signs that say No Swimming and No Fishing. I visited on Buddha's birthday, and one fly angler was there, along with a spin angler fishing in a pool near the trout farm. There was a much-cluttered feel to the valley. Also there was a tourist information booth, which implied that it gets hit by many visitors. I never returned.

In the Yeongwol area, I fished Jikdong Valley near Kim Sat-gat's old residence. He was a vagabond poet, a kind of Korean Walt Whitman. There were more commercial fish farms and escaped feral rainbows. No other anglers were there, and the fish population was spotty. I caught a couple around twelve inches. The valley was nearly perfect for guiding. It was remote and scenic, and the village had an old-time hamlet feel about it. The creek ran near the road, but there were plenty of places where it diverged. A huge number of case-building caddis larvae speckled the underwater rocks, but the hatch had not started and no flies were airborne yet.

I edged closer to fulfilling a dream: to write for *National Geographic* magazine. I grew up reading it and dreaming of adventure and mystery in distant lands, and that reading habit most likely set my path in life. I got my foot in the door with National Geographic News, an online bureau that published dispatches from around the globe. I wrote a story about a forensic archaeology team, with the U.S. Joint POW/MIA Accounting Command, who were in country to find the remains of an American soldier.

After that, I pitched a story about a salmon restoration project in South Korea. I had double-dipping motives: write the story and catch a chum salmon, on the side, with a spey rod. My years of studying kumdo swords-manship came back to me. The two-handed grip on the spey rod was similar to holding the bamboo sword, and the flex and application of power were nearly identical. I loved the athletic elegance of two-handed casting and I hooked into a half-spent salmon almost two feet long, but I did not like what I learned during my reporting.

There was not much if any restoration going on. The Namdae River, which the salmon migrate into, had plenty of dams and weirs to impede their journey. A huge net stretched across the river, and the salmon were captured and sent up a conveyor belt to a bankside fish hatchery. They were milked, and their milt and eggs were fertilized in tubs. There was little, if any, natural reproduction in the natal water. They were merely gaming the salmon's homing instinct: release the artificially spawned salmon fry and, years later, mature salmon return. It was like releasing a calf born with a homing device around its neck and letting it go to graze in the commons for free, and it returns years later plump and ready for the butcher's knife. The nearby town of Yangyang held an annual salmon festival to celebrate this wonder, and the organizers fenced off a shallow section in the Namdae for tourists to splash in and catch trapped salmon with their bare hands. I left this shit show disgusted and drove upstream, past more dams and weirs and deeper into the mountains. I fished the Bopsoo branch and caught native stream–born cherry trout. I later sold the spey rod to a client and broke even.

Fig. 18. Other than a few stray sea-run cherry trout, dog salmon, *Oncorhynchus keta,* are South Korea's only anadromous salmonid. Reproduction is mostly artificial. Photo by James Card.

There were many streams that I fished once or twice, and I had to make judgment calls on the stream and its fishing quality and its surroundings. Was it worth another long drive? Was it worth another weekend better spent at another locale? For some creeks the answer was no. All of these places were in Gangwon, South Korea's northernmost province, cold and rugged, and even the diet of these mountain people is changed as there are few rice paddies. Outsiders call them potato eaters.

There were many dead ends. I fished Nakpoong Creek in Okgye Valley. Two streams flowed into a reservoir. The right branch was shorter and ran through a village. The one on the left ran over three miles into the backcountry. There was a trout farm where many rainbows once escaped, and past the farm the valley was rugged and held stunning water. I talked to the trout farmer. He said the fishing in the valley wasn't good and I should head to the reservoir—which, I had noticed on the way in, had No Fishing signs posted. I hiked up the valley, casting into each pool, run, and riffle.

Thousands of black caddisflies flittered over the stream. I cast black caddis patterns for the length of the hike and caught only a couple of chubs. No trout strikes at all. No trout spotted and no telltale darting shadows. The stream was exceptionally clear and perhaps they were lying low for the day, but that was not a proper excuse. The stream was characterized by many limestone flats, or steps, which were pretty to look at but held no fish. There were a couple of abandoned farmhouses with collapsed roofs, along with stone wall ruins. The hillsides were wooded, with a few sections of secondary growth from an old forest fire. Other sections of the mountainside were fallow farm fields. I found a poacher's two-handled dip net lying on a rock midstream. At a clearing near a bamboo grove, I came across an offering to San-shin set upon an altar flatrock. It was a remote valley and roadless area and had the potential to be the top trout stream in the country, yet not a single trout rise. After exploring the left fork, I explored the right fork. It was another cement-sided river, dotted with many small Korean farms. The stream was nothing more than a trickle.

There's an old saying—if the mountains of Korea were flattened, the country would be as large as China. It was the mountains that provided so many trout streams, but it was the mountains that added to the driving time. I was always doubling back, up into a hidden valley, back down again, and then up and over another mountain and back into another valley. There was so much to explore. I thought many times about getting a new job up north and moving closer, but it wasn't so easy with my wife being a public school teacher. Moving to a different school was like a soldier putting in a transfer to another unit, a complicated process involving an earned-point system. I was tired of the vague paranoia of going a little too far into the *mintongseon* and getting in trouble with some soldiers. I needed to find lenok water closer to home. My first focus had been heading up north. Now my focus needed to be down south.

10 The River of Records

The world's southernmost population of lenok is located in a remote sub-range of mountains where the landscape is darker and a few degrees cooler than other parts of the country. It is as if the mountains oozed a cold sweat under rupestral mosses and the trees threw more shade. In my office I had a huge topographical map of the South Korean Peninsula taped to an enormous sheet of stiff cardboard. I could move it to the kitchen table and lay it flat and stand over it. Usually, it leaned on one of my utility tables, and from my desk I could glance over and plan the next trip. On it were secret codes marking prime fishing spots throughout the country.

There was one remote pocket of territory that was the blankest spot on the map. My eyes were always drawn to it. I passed by this region many times on my way to the DMZ country. Not much was there. There are few main highways. There are more tiny villages than main cities.

This remote region is part of the Baekdu-daegan, the mountainous backbone of Korea. It is the country's great watershed and birther of its rivers. It is the nation-spanning wildlife corridor that spreads rumors of the modern existence of leopards and tigers. It holds thousands of coulees and micro valleys—many untouched and undeveloped. The Baekdu-daegan is considered Korea's spiritual spine, and it creates a life force so strong that Japanese colonialists pounded metal spikes into its peaks and ridges to sever the flow of national energy to cripple the Korean people.

It makes sense that this remote area is where the Joseon Dynasty hid its most valuable treasure. The Annals of Joseon Dynasty are records from 1392 to 1863. The annals comprise 1,893 volumes. For a long time, these were written in classical Chinese calligraphy, but during the rule of King Sejong (1418–1450) they were printed in movable metal and wooden type. Only later, in the 1980s and 1990s, were they translated into modern Korean.

The four libraries where the copies were stored were scattered throughout the peninsula. During the Japanese invasions of 1592 to 1598, the Japanese burned down the repositories in Seoul, Chungju, and Seongju. The only repository to survive was the one in Jeonju. Five more copies were printed after the war. Most were stored in Seoul, and the rest were sent to the safest place in the country: the blank spot on my map, the rugged mountains of Taebaek and Bonghwa Counties.

It is part of the country that is always cold. The snow stays longer in the winter, the trees change colors earlier in the fall, and spring is always late to bloom. It is said that from ancient times this area has had no wars, disasters, famines, or plagues. The mountains here are not as glamorous and photogenic as those in Soraksan National Park, nor are they as expansive and hiker friendly as those in Jirisan National Park. These mountains are composed of steep dark-colored rock faces, and the tightly walled valleys are claustrophobic. The forest is dense and offers little for distant vistas. There are dark, mysterious holes in the earth far up on the mountain slopes. There are many abandoned anthracite coal mines in this area. The zinc mine on Yeonhwa hill was the largest in the country and closed in 1992. Over two hundred families once lived here, and the buildings are empty. There are old logging roads and overgrown and collapsed *chogajips*—houses with straw-thatched roofs—with crumbling red-clay walls. Some of the old folks came to these deep valleys to escape the ravages of the Korean War, and they never left.

Two miracles occurred on a countrywide scale in South Korea. The first and most publicized is the Miracle on the Han River, which refers to the nation-building chapter in South Korean history—going from being one of the poorest countries in the world after the Korean War to emerging as a nation prosperous enough to host the 1988 Summer Olympics.

The other miracle is more mundane, but it allowed me to go fishing almost anywhere I wanted. South Korea is a heavily forested country, but at one time it was a very barren land. I loved to study old black-and-white photos from the turn of the century to the time of the Korean War. In the images I look past the people and the buildings and focus on the hills in the background. They are all treeless. They have been picked clean of every stick of firewood. It wasn't until the turn of the nineteenth century that the *yonton* was introduced and used as a heat source. The *yonton* was a large briquette molded out of coal dust and perforated with twenty-two holes. These were used to heat homes, well into the 1980s, until gas and oil furnaces were adopted. It was during these decades that the woodlands could recover from the firewood scavenging, but also the government launched a massive reforestation effort in the 1970s. To this day, the timber is still young, and South Korea is not self-sufficient with wood products and must import lumber.

These wooded areas are simply known as the "greenbelt," and they were the reason I was confused on my very first hike on Mireuk Mountain: I thought I was trespassing on somebody's land. The greenbelt policy was enacted in 1971 to protect undeveloped land, wilderness areas, and farmland near urban areas. It's an invisible line, a national zoning code that allowed the wild areas to regenerate. There is much private land within this zone. My wife's family owns half of a small mountain within the greenbelt. Other than the graves of a few ancestors, not much is done with it, so nobody cares if a hiker passes through that greenbelt-zoned land. Other than some small orchards and vegetable gardens in the foothills, the land is left in a re-wilding state. A year later, after the greenbelt policy, the River Act of 1972 nationalized all land along the nation's rivers. This instantly turned every inch of riverfront land into public access. Once I discovered this, I realized I could access nearly any river system anywhere I wanted as long as I was willing to hike for it and politely detour around anyone's gardening activities. Other than stomping through a farmer's rice paddy, entering a secured area, or climbing over a courtyard wall, the concept of trespassing is very different than it is in America. Anglers in the U.S. are restricted to access points along a river—whether it is via public land or by permission from private landowners. In South Korea, I did not have these concerns. The country was wide open for exploration except for a few places.

There was a sign near Baekchon Valley that stated that fishing for lenok in the valley was prohibited. Manchurian trout in Daehyeon Village of the Seokpo Township in Bonghwa County were protected and the fish was designated as Natural Monument No. 74. There was no mention of where those boundaries start or end and no reference to catch-and-release fishing, because that was a foreign concept. When I hiked these angling-prohibited areas, I always found monofilament tangled on driftwood snags and over-hanging tree branches. The locals were certainly fishing for them. I avoided that area and fished other places. National parks seemed to have similar fishing prohibitions. I hadn't set foot inside any of the national parks in years. Instead I worked the edges and boundary lands, and I did the same in the Taebaek-Bonghwa country.

The lenok fishing in these remote cold-air valleys was good, and I started guiding trips in the area. South Korea is a net exporter of tourists. Many Koreans were heading out and fewer coming in. I know this because many of my fly-fishing clients were not tourists but happened to be in South Korea for other reasons. This glaring trade surplus grinds the bureaucrats, so they hire PR firms to lobby international groups to host events so people are forced to visit.

I guided an elderly British gentleman who was the last man standing in his tontine. He and four of his fishing partners had each contributed 5,000 Euros to a prize awarded to the man who lived the longest. The winner was to put the sum toward angling adventures. South Korea was a layover stop when fishing for Mongolian taimen. He looked me up and booked a trip and decided to stay a couple of days. Another layover client was a Texan deminer that worked for an NGO removing unexploded bombs in the Plain of Jars in Laos. He asked me many questions about Asian women because he was in love with a Black Tai woman from there and wanted to marry her and bring her to his hometown of Lampasas. In return, I asked him many questions about Guadalupe bass. I guided an American missionary en route to Papua New Guinea, and when he caught his first lenok, from a creek shaded with trees turning their autumn colors, he said he wanted to say a prayer. He said this place was so beautiful that it was very close to heaven. I agreed and we bowed our heads.

Henry David Thoreau wrote, "Many go fishing all their lives without knowing that it is not fish they are after." Thoreau was a poor fisherman, so his platitude on the subject doesn't count. From what I have seen as a fly-fishing guide, his line could be rewritten as: many go fishing all their lives without knowing that they are trying to recapture childhood memories or ones they never had. This thought applies only to the small streams

Fig. 19. The southernmost population of lenok live in a handful of creeks in the Taebaek Mountains.

where I guide clients. As boys they scrambled through creeks and canals and rivulets and flooded ditches, getting soaked and muddy. They explored farther and farther upstream to the source, fascinated by flowing water. Along the way they caught frogs, turtles, and minnows using string and sticks and salvaged flotsam. As adults we recreate this experience in a more mature way, using a semi-technical system (fly fishing) to wade and muck in and around small streams to catch a small but very pretty fish and reduce it to a handheld possession for a moment.

I parked off the road down from the temple before sunrise. I hiked past the *jangseung* totem poles (also known as "devil posts"), placed long ago to fend away evil spirits and supernatural forces bringing infectious diseases. They were my favorite kind: an upended pine tree with the entire root system reaching toward the sky, and carved into the trunk was a buck-toothed, bulgy-eyed guardian, the tree roots forming its wild hair. I was on the creek as the sun was coming up. The tea-colored water streaming out of the chute reminded me of the poured beer I drank the night before. The Korean equivalent of pale morning duns were on the riffles like wisps of smoke on the water. I tied on a small cream-colored mayfly and smeared on a speck of float grease. These waters are always so dark and shaded that I was always happy to fish with a light-colored fly. My favorite blackened deer-hair caddisflies were always tough to keep an eye on as they floated over riffles, to the point that it was mentally exhausting to fish with them. I preferred bushy, high-riding dry flies, brightly colored ones, on these dark waters, and I trimmed the bottom hackles for a lower profile and a more natural appearance.

I tossed out the pale fly and bang! Instant strike. This is why you get out of bed with a hangover. Life rewards those who show up. How you feel at the time of showing up is irrelevant, but you have to show up and put forth effort. It was a nine-inch lenok that jumped three times. I stepped back at the end of the pool to release the fish, and it hung in the shallow for a second before it darted into deep water. More lenok rose at the head of the pool, riseforms polka-dotting the water. I cast again and caught another about the same size.

This is a very curious conundrum for small-stream anglers. It is about greed: How many fish can be caught out of a pool before the fishing quiets down and the fish no longer bite? The answer is that it depends on the pool and the nature of the stream. In the clear-water pools of the Jirisan streams, once a trout is hooked and it is diving, running, slashing, splashing—the other trout in the pool freak out as if a rock had been tossed into the pool, and they vanish to hide in deep crevices. Yet on other streams, especially these dark-water freestone streams with a forest canopy and long sections of fast water, numerous trout can be picked off from one casting position. My record was seventeen, at a long stretch of riffles that bent around a huge boulder. The fish forward of the boulder did not know that the ones around the backside of the boulder were getting steadily picked off. The aggressiveness of the feeding frenzy during a hatch is another factor. Sometimes the fish let their guard down for the reward of claiming easy calories. It is the same for all animals. It's the whitetail deer that steps out from the safety of the woods and into an open field to gorge on soybeans as a hunter pulls back his bowstring.

I caught another lenok and moved ahead. That was enough for that pool. At the next upstream pool there was a break in the tree canopy, and warm morning sunlight poured into the pool. Two Korean pines that had once shaded this spot, but had been uprooted and had fallen into the creek, added to its character. The current flowed under, above, and around the obstacle and scoured away new spaces underwater. The trees were long dead and no longer held their bark.

There was no insect activity on the sunlit pool, yet small pale mayflies hatched on the pool below it in the cold, dark shade. The cooler pool was hot; the warmer pool was not. I watched the water and was able to look through it. If you watch the surface of riffled water, you will eventually see what I call a "windowpane" of clear water. It's a riffle-free slick that passes over a deep spot or a smooth object and allows you to peer, for a split second, to the depths of the pool and catch a glimpse of a trout. Windowpanes do not appear at all spots, and they come and go depending on rainfall, wind, and water velocity. Through a windowpane that passed over a spot next to one of the fallen pines, I saw the sweeping tail of a large lenok.

I snipped off the pale mayfly and tied on a Hornberg, a fly created by Wisconsin angler Frank Hornberg in the 1920s. It's a large dry fly that imitates a stonefly or grasshopper, and at the end of the drift, it can be yanked underwater and retrieved to imitate a minnow. Since nothing was hatching here, I would throw this fish some meat. I waded as close as possible and short-lined the pool, by flicking a leader's length of fly line upstream, and lifted the rod as soon as the Hornberg flicked onto the water. I pulled out the slack and walked the line above the bubble line tailing from the fallen pines.

The lenok smacked the fly hard like a shark taking out a swimmer at the beach. It jumped halfway out of the water and dove down under the pines. I kept the line tight on him and wore him down with small strips of line, an inch at a time, as I waded closer, until he was almost under my feet. I dunked my net under the pine log and captured him after a minute of blind groping and scooping. He measured fourteen inches against my rod markings, a trophy for a creek this size. I slipped him back under the pines.

I spent more and more time up here because my driving time was minimized. I could make it to this region in a little over three hours—compared to the day-long or overnight drive to the streams near the DMZ, which left me frantic to maximize my time on the water. My expenses in gasoline and tolls were halved too. For a while, to save money, I was buying Cenox, a bootleg fuel mixture that was half the price of regular gasoline because it was tax-free. Cenox was the subject of a long-running legal battle as the maker claimed it was a "fuel additive" and not a fuel. The Supreme Court eventually ruled in favor of the petro producers and deemed it an "imitation gasoline" and banned for manufacture and sale.

On a side street in my Changwon neighborhood, there was a cardboard sign duct-taped to a cement wall. It said to drive to a gravel parking lot two blocks away and then call a cell phone number. Upon calling the Cenox dealer, I was told to park near a box truck with a flat tire. A minute later he drove up on a yellow scooter. He unlocked the cargo hold of the truck, and inside, stacked from floor to ceiling, were eighteen-liter tin jugs. He popped off the plastic lid, attached a nozzle, and poured the automotive moonshine into my gas tank. The liquid had a sweet smell.

To get up here I drove north and, as I passed Daegu, I tuned into the Eagle 88.5 to see what was happening on Armed Forces Radio. Daegu is the country's third largest city, a drab metropolis that is an urban heat sink with canyons of identical apartment blocks as attractive as grain elevators. It is the center of the textile industry, so there were some public relations efforts to rebrand the city as the "Milan of Asia." That did not work out. It is famous for apples, and it is said the most beautiful of all Korean women come from Daegu, so at least they have that going for them. Once I got past the gladiatorial driving around Daegu, it was an easy cruise.

It was a fine time to be in the woods. Autumn is called *cheongomabi*, meaning in Korea, the "season of high skies and fat horses." The sky was clear and bright blue and brought a breeze that was warm and cool at the same time. It was the harvest season, and the farmers' markets were loaded up with fresh produce. It would soon be the time of *kimjang*, the annual kimchi-making season, when housewives stocked up on radishes and cabbages and put up a year's worth of kimchi, stowed away in brown ceramic pots for fermentation. *Chong-gak kimchi* (bachelor or ponytail kimchi) was a staple on my river rambles, and I kept the spicy chunks of crunchy white radish in a small jar in my fly vest.

With this surplus amount of time, I was able to scout and hike more and take it all in at a slower pace. I ditched my fly vest and leaned my rod in the crook of a boulder. The pine mushrooms were emerging, and it would be a shame not to look for some. They appear in the places that have Korea's highest concentration of freestone cold-water trout streams—in the Gangwon and the North Gyeongsang provinces.

They are better known in America by their Japanese name, *matsutake*, which translates to "pine mushroom." The Korean name is *songi beoseot*, which also translates to "pine mushroom." Hunting for pine mushrooms reminded me of hunting morels in Wisconsin. They have two similar characteristics: there is only a month-long window of opportunity to harvest them, and then they are gone. The other is that both types of mushrooms cannot be commercially cultivated. Agricultural science has not figured out this puzzle, but if it is ever solved it will be a billion-dollar enterprise. Until then you have to hunt for them. Hunting for them is difficult and time consuming and also a crapshoot. That means they are some of the most expensive and desired mushrooms in the world. They have been called the "truffles of Asia."

I crossed the creek to a stand of older red pines (*Pinus densilfora*) on a steep slope with good drainage. By looking uphill at an angle, you can better see the white stems. When searching for pineys on flat land and looking straight downward, you see the brown tops, and they are harder to spot on the forest floor.

The other thing that I kept an eye out for was wild ginseng. Wild ginseng hunters known as *shimmani* hold ceremonies to conjure the mountain spirit to guide them to the root. Often, they are guided by a dream the night before. I never found any but if I did, it would be a life-changing event. A wild root can easily put ten or twenty or thirty thousand dollars in your pocket. The older the root, the more valuable and potent it becomes. Roots over a hundred years old have sold for six figures at auctions and make headlines in newspapers. Wild ginseng is considered to be more healthful than the cultivated variety, which is sold fresh in supermarkets. I often ate the cheap farm-raised ones, and I would keep a root in my shirt pocket and chew on it throughout the day. It tasted like a bitter, earthy carrot.

I searched for them for half an hour, found nothing, and returned to fishing. I close-worked a couple of small pools and riffles and crawled and kneeled to casting positions. I caught a lenok, and when I released the fish I noticed an odd-shaped rock. It was black and squared off and cracked. I picked it up, and carved into the stone was a three-legged crow, a *samjoko*, a powerful symbol in Korean mythology. Part of the stone had a smooth indentation. I had seen it before. It was an inkstone. The aristocrats decorated their inkstones with strong mojo. They often used the ten symbols of longevity: sun, clouds, water, rocks, turtle, crane, deer, pine tree, bamboo, or the mushroom of immortality (the *Ganoderma* family).

Where I released the lenok was a flat-topped boulder at about chest height. I climbed on top of it and looked upstream and downstream. Upstream was the prettier view. I sat down cross-legged and imagined parchment spread before me and pretended to swab an imaginary brush on an inkstone. I placed my hand on that spot and leaned over the edge of the boulder and looked down. It was where I found the broken piece.

Satisfied with my conclusion, I scaled down the boulder and fetched my fly rod, and then I saw the mushrooms. I bear crawled and grabbed pine trunks to pull me up the steep slope. There were five. They were about three inches long, the heads had a muddled brownish color, and the stems were white—and it was the white that caught my eye. I pinched them off and tucked them into my wader pocket but not before I spotted a *jinae* slipping under a rock. It was a common venomous centipede (*Scolopendra subspinipes mutilans*), which can grow to be up to seven inches long. I was bitten once when I lived in Tongyeong, and my hand swelled up until my fingers were like bratwurst. Its venom is a histamine-like chemical similar to a bee sting. Herbal medicine shops in the old market districts sell them in dried bundles.

I hiked farther upstream, walking more and fishing less and making only quick casts into the most promising riffles. The lenok became smaller and so did the water. I arrived at a helicopter landing pad deep in the woods. It was for either the military or the Forest Aviation Office, a branch of the Forest Service organized to fight wildfires. There wasn't much of a military presence in this region, unlike the streams near the DMZ, and that was a relief—the freedom of not having to look over my shoulder and not having paranoid thoughts of stepping on a landmine that had been washed downstream by monsoon floods.

I achieved my goal—to be able to fish for lenok closer to home. It was the world's southernmost population of this ancient cold-water species, but my plans changed slightly. I would head back north, just a little, because the next valley over was so geologically glorious. It was a great watershed of limestone that gushed with artesian water. I hiked back to the car early to give myself time to make it over the mountain pass.

11 Into the Karst Kingdom

The best fly-fishing wisdom I've ever come across is from Tom Rosenbauer, who wrote that geology determines the character of a trout stream. The geology will predict the size of the trout, their population and distribution, and their forage. Your angling possibilities were determined thousands of years before you even found the stream on the map. This is what I thought about as I explored the Joseon Supergroup, a jagged mountainous area overlying Precambrian granitic gneiss and layered with limestone, sandstone, and shale, which covers Samcheok, Taebaek, Yeongwol, Pyeongchang, and Mungyeong Counties.

From the Taebaek region, I drove over a high mountain pass, on a series of switchbacks, and descended into the Osip River valley. It is the longest trout stream that flows to the coast, and it empties into the East Sea near the city of Samcheok. The name "Osip" means "fifty" as it bends that many times through Korea's largest limestone region before reaching the saltwater.

Samcheok is known for cocks and caves. The Haesindong Shrine is an outdoor exhibit of wooden phalluses; some are huge, some are plain, some are intricately carved. There are penis totem poles, an enormous cock cannon on wheels, and hard-dick park benches to sit upon, along with a dildo-riding seesaw. There are statues of ancient fishermen holding their erect shafts pointed to the sea. It is a popular place for female tourists to pose for raunchy photos. The shrine was founded from an ancient tale of a virgin who drowned while gathering seaweed. Bad luck befell Samcheok's fishermen, and the villagers made the phallic shrine to appease the spirit of the virgin who died before she could ever experience the pleasures of getting laid. They have been carving wooden dongs ever since.

I came for the caves although I did find a driftwood log sculpted smooth by water. The wood grains resembled swollen veins, and one end was a bulbous knot. I threw this priapic objet trouvé over my shoulder and hiked it back to the car. When I drove into Samcheok later in the day, I left it leaning against a pine tree at the shrine.

The Samcheok region has the highest concentration of limestone caves in the country, around fifty-five, with more probably still undiscovered as the mountains are so steep and rugged that they discourage exploration and excavation. The largest one, the Hwanseon Cave, is located in the Daei Valley, about thirty minutes outside of Samcheok. Daei Creek was once home to a rainbow trout hatchery that was blown apart by floods long ago, and that event populated the Osip watershed with rainbows. The crumbled concrete raceways still exist. To generate tourist interest in the caves, the city of Samcheok erected one of the ugliest buildings ever created in Asia. Called the "Hall of Mysterious Caves," the structure looks like a wedding cake dripping with diarrhea. The icicles of shit are supposed to be stalagmites. Inside the building is a fake limestone cave, an IMAX theater, and paleontology and speleology displays.

I descended into the valley on a series of switchbacks, and on one blind corner a Daewoo Tico was crumpled into a concrete barrier, which prevented the vehicle from plunging into the canyon. The doors were compacted and covered with blood, and the small car was empty of passengers. I passed Dogye, a decrepit coal-mining town built on a high slope in the upper valley. I looked out of the passenger-side window at flashes of the river far below. From the headwaters near Dogye, the river flowed about fifteen miles into the city of Samcheok as the bird flies. The river was probably double or triple that length if the fifty bends are factored in.

The Osip was populated with rainbows and cherry trout and the rainbows were overtaking the cherry trout, and I thought there were more stable populations of cherries upstream. While casting around for these two, I discovered another marvel of Korean native fish, the Pacific redfin. It is a cypriniforme that behaves like a salmon or sea-run trout. In the spring months, they migrate upstream in rivers to spawn. There are two species, *Tribolodon brandti* and the big-scaled redfin, *Tribolodon hakonensis*. Koreans

Fig. 20. The Pacific redfin, *Tribolodon brandtii*, is a sea-run carp that is similar to a trout. Photo by James Card.

call the fish *hwangeo*. As the name suggests, their pectoral, pelvic, and anal fins are tinted red. The big-scaled redfin has a crimson nuptial line running along the white of its belly, while the *brandti* subspecies tends to be less colorful.

They are members of the Far Eastern daces, a genus with only four species of its kind, which evolved around rivers that flowed into the East Sea. The other two species are found in Japan and Sakhalin. They grow to sizes similar to small-stream trout, with smaller ones around eight inches and trophy-sized ones around eighteen inches. Their feeding patterns and behavior are nearly identical to trout, although I found subsurface presentations worked the best.

The most interesting thing about their anadromy is that, like trout, not all of them migrate. Some go out to sea, and a few others are river residents. Although they show up in greater numbers during the spring, I caught them on occasion throughout the year. The only explanation I had for this mystery was the presence of numerous low-head dams throughout the length of the Osip that thwarted their migration.

The goal was to catch all three species on every trip into the valley. I found the redfins to be in the lower and middle stretches, the rainbows in the middle, and the cherry trout surviving in smaller pockets farther upstream. I thought of Robin Vannote's river continuum concept, that rivers must be understood as a continuum linked to a greater watershed, not studied as static slivers of water. It mapped out the sequences of biological communities throughout the course of a river. It is an elegant and scientific description that explains why trout are found far upstream and catfish are found way downstream. There was one problem with the concept, and that was the presence of disruptions to the flow of the river. The numerous dams on the Osip throttled the river continuum concept. The river was now a mutated and disfigured continuum, and, from dam to dam, the river took on different characteristics as if I had to study each section as its own stand-alone ecosystem. To the untrained eye, it looked all the same. To a curious angler, it was an ugly and unnecessary inquest I did not ask to solve.

Writing about Korean rivers at this moment in history is like prepping an obituary of someone about to die. South Koreans alive today have lived with unnaturally flowing rivers every day of their lives. They never saw it coming. It is the ratcheting effect of poured concrete; there is no going back without great effort. The mission creep of the cement industry: to line every creek with concrete and amputate the flow with dams and weirs. There are over 18,000 dams in a country the size of Indiana. It's riverine acrotomophilia. It's a forlorn hope.

Korea was once called *guemsu-gangsan*, or the "beautiful land of embroidered rivers and mountains," but in many watersheds it is no longer like that. It is the land of stunted rivers with straightened, meanderless channels, and the Osip was my case study for how much abuse a river can endure and still sustain trout. Its treeless shores are covered with miles of cement riprap like plaqued arteries. Wired-up gabions bung up the banks, and there are ejaculations of shotcrete where the cement trucks can not reach. In some spots poured cement walls, like that of a building, create a concrete canyon effect. This, in turn, causes more channelizing downstream, and more concrete creates more concrete. The channelizing creates more

Fig. 21. Invasive non-native rainbow trout, *Oncorhynchus mykiss*, escaped from commercial fish farms, formed wild populations, and compete with native cherry trout. Photo by James Card.

current velocity during heavy rains, and those downstream take the brunt of the river's power during the flood season.

Excavators scraped the riverbed flat for aggregate, erasing the thalweg. Like a blood disease, runoff has infected the now-shallow and now-heated stream water, and algae blooms are suffocating underwater life. The Osip flows in quiet desperation. Few South Korean children will know what a natural river looks like unless they travel overseas.

What keeps the Osip alive and wild are the bends that give it its name. The bends bang up against miles of sheer cliffs, and it is these places that the earth-diggers and concrete-pourers left alone. To face a riffled run that flows tightly against a hundred-foot limestone face is a wondrous moment, but turn around and you will be looking at a concrete embankment and more land sculpted by man and machine. In many ways the Osip was like the tumbled-boulder streams of Jirisan, in that an angler could predict catching fish in finite, countable spots. But whereas the Jirisan streams were broken up by naturally occurring giant boulders and waterfalls on a very steep gradient, the Osip was broken up by low-head dams, streambed scouring, and channelization.

The Osip had so many newly built redundant bridges to nowhere that my atlas was outdated. My atlas of Korea no longer illustrated the true nature of the landscape. The thin blue lines denoting streams had been reduced to sclerotic trickles within cement walls. By following the watershed drainage pattern on a topo map, the spider web tendrils of tributaries, large and small, nearly everywhere you'll find the same situation: cement. This is the most devious of all environmental destruction—the small stuff that people notice but do nothing about. Korean environmentalists have vigorously opposed the building of large multipurpose dams in the past, and there is outcry when the government oversteps spending on huge boondoggles. There are few natural lakes in Korea, only impoundments. But it's the small low-head dams on small rivers—that are obsolete as soon as they are built—and channelizing the banks and the unnecessary bridges that nobody questions. The construction industry flogs unnecessary development as a way to improve rural communities.

They are Frankenstein rivers and lack oxbows, meanders, and sloughs and are ecologically crippled. The channelization of the rivers is a damaging and obtuse way of dealing with flood control. The riverbed is dug out and the benthos and hyporheic zones are gutted, and as if that weren't enough, the riverbanks are defoliated of riparian plant life and the bio-productivity is degraded. Recovery is hard. After the embanking and channelization of the river, the riverside is barren until pioneer species start growing. There are no trees, there is no shade. Weeds take over at best. It is difficult to have regrowth since little soil accumulates between the cement riprap, and few seeds are left in the soil to repopulate the earth. The important relationship between the river and the forest is shattered. This means you are fishing in the sun. The only shade trees are ones growing on the cliffs and upper slopes. There is no canopy cover. It is best to fish the dark shadow water under the cliffs. Since the soil, brush, and trees have been removed from the riverbanks, runoff occurs. The surface water from rainfall isn't absorbed and held in the soil. Rather, it runs over the nonabsorbent surface at a faster and more polluted rate causing flash floods downstream.

The ecotone is the transition zone of ecosystems. It's where the forest turns to field; it's where wetland becomes solid ground, where the foothill meets mountain. Native Americans thought these to be places of power. Wise hunters always stalk the edge. In Korea, the ecotone is of cement meeting water. There are no riparian buffer zones. A flooded river is always looking for its natural floodplain. Floodplains are flat and in South Korea, where flat land is rare, the floodplain must be reduced to a possession and turned into sellable land by separating it from the river via a concrete barrier.

The hydrologic action of these streams is often reduced to a trickle in times of drought. With much of the riparian area covered in cement, much of the rainwater runs off into the stream rather than permeating the soil and recharging the water table. This has created ephemeral and intermittent streams: streambeds that have water flowage during times of rainfall or in the rainy season. The rest of the year, they are dry. There are some fish-passage ladders on the Osip, and usually they have no water running over them. They are nothing more than a creative excuse to use more concrete. South Korea is one of the largest cement-producing nations in the world, and the cement pouring is an ice-nine and gray goo scenario out of control.

Dam building and river channelization for the excuse of flood control is the self-licking ice cream cone for the Korea Water Resources Corporation, a bureaucracy that pathologically overuses the hollow phrase "eco-friendly" in all of their literature about dam building. Every project is colored anodyne green. Paving rural areas in concrete creates jobs, and I don't blame the country folk for taking such jobs. It's either that or depart this gorgeous valley and move to a dreary industrial city for work. Because Seoul hogs the nation's wealth, South Korea has no-poor-province-left-behind policies and engages in make-work projects in the hinterlands to keep the hillbillies from grabbing their pitchforks. Many rural folks welcome the building of dams because it is their ticket out of poverty. The government will compensate them for the farmland that gets flooded, and they can start anew somewhere else, flush with cash.

If rivers are the arteries of an ecosystem, the creeks are its capillaries. The surrounding landscape of the karst kingdom bleeds clean, cold water.

Rills trickle off high cliffs, and rivulets cascade through limestone gullies. The newest cave to be discovered was Daegeum Cave on Deokhang Mountain. It was opened for public tours in 2007. It has stalactites, stalagmites, and large waterfalls. The karst topography of the area is Mother Nature's plumbing system for trout streams. I had fished the entire length of the Osip many times. Now it was time to investigate the tiny feeder creeks and steephead valleys that were unblemished by concrete. Sometimes one must go farther afield to find love outside the ruins.

A side valley looked promising on my topographical map because it was on the opposite side of the Osip in a spot inaccessible to earthmovers or other vehicles, and the Chosung Cave was marked within the valley. I left my travel fly rod cased up in my backpack and waded across the river to the mouth of the valley, which was more of a steep-angled ravine. I hiked up into the riverbed. It was dry. There was no moving water, only puddles in low spots—perhaps from the last rainfall. It was evident the streambed had been sculpted for thousands of years by water. The space left by the limestone gouged out of the canyon was large enough to drive a jeep through it.

I hiked a mile up the streambed, which was shaded by trees on both sides. If it had been flooded waist deep with cold water, it would have been the most perfect trout stream in the country. This was the trout stream I'd spent years looking for, and it held no water. I hiked farther upstream in the nonexistent stream and sat down on a smooth limestone ledge and thought of Nestlé, the Swiss hot cocoa food conglomerate that got into the bottled spring water hustle. The company had past trouble in the United States for sucking aquifers with a vampiric rapacity. Perhaps somewhere on this mountain was a similar pumping station to feed South Koreans' love for bottled water, because no one trusted their tap water, and the action bled this creek dry. There could be lateral erosion, where one stream breaks into a nearby stream and starts flowing into it, leaving its original course dry. That did not make sense. The valley was too steep for the creek to flow anywhere but down. The answer I settled on was the creek disappeared underground and reappeared God-knows-where: in an underwater spring in the Osip riverbed; in another cave, discovered or undiscovered; in another feeder creek; or into a great underground river. The answer would be farther up the streambed, but that would have to wait. The day was slipping by, and I had not caught a trout yet.

Fig. 22. The channelization and damming of the Osip River was brutal, yet stretches of the river produced rainbows, cherries, and redfins. Photo by James Card.

Back in the car I drove back down the Osip and cut over to another side valley. I drove past a spot where I had guided a client a month ago. He lived in Osan and was following in the footsteps of Chuck Norris; he was studying Tang Soo Do. "So there's trout in here, eh?" he said. He did not seem impressed by the beautiful view of the river running against the cliffs. I told him there were rainbows and cherry trout in the Osip. As if on cue, an osprey plunged from the sky, smashed the water, and heavy-winged it back into the air with a fat rainbow. The client scrambled out of the car and humped himself into a pair of waders.

When I first investigated this creek and peered at it from above on the bridge, it was nothing more than a skinny stream with no cover or holding water. I got out and hiked upstream anyway and found two clues: I spotted a plank of twisted Marston matting wedged into a rock—a relic from the Korean War. Cherry trout fry swam through its perforations. Above that was a larger pool, and I found a broken bamboo pole tipped with monofilament.

Beyond that were a few riffles, small pools, and glides, and I caught a few cherry trout from those spots. I was very happy that the anemic creek at the bridge had turned into a small but fishable stream. I hiked farther upstream, and that's when I found the Big Hole. That's what I called it. There is a famous trout river in Montana of the same name, but that is unrelated. It simply was a very big hole, the largest I had ever seen. It was circular in shape and about twenty feet deep, maybe more. The water was clear with a touch of turquoise. The creek plunged down a waterfall chute into a small, curved upper pool. Then it poured over a lip into the Big Hole, creating a frothy bubble line. The waterfall drops formed the front of the Big Hole, the sides were limestone walls, and the rear of the hole was sheer underwater wall. Water flowed over this hole onto flat-rock shelf and then downstream, forming a normal-looking creek.

I caught some nice cherry trout on that first day. On my return, I brought my wet suit, mask, and snorkel. I edged along the top of the limestone wall and body-skidded down into the upper plunge pool. Then I eased into the gelid water, gripping the rock against the force of the current. There they were: a school of small cherry trout floating up and down and nipping at food flushed down in the frothing water. I watched them as long as I could—then I lost my hold. The force of the current knocked me backward, and I somersaulted over the lip and plunged into the Big Hole. I surfaced and gulped air from the shock and treaded water for a moment and let the current push me against the limestone wall. I hung there for a while to let my splashy entrance into the pool die down. I lowered my mask and snorkel underwater and looked below. It was an aquarium of cherry trout. All healthy and of all sizes—then I spotted him, far below. I nicknamed him the Salmon, and he was the biggest cherry trout I'd ever seen, around twenty-four inches. That fish swam in the Big Hole like a haunted submarine. All other fish avoided him. He was a cannibal.

On subsequent trips I threw everything in my fly box into catching the Salmon. The problem was the Big Hole was too deep. I used sinking-line and split-shot and weighted flies and hybrid fly-like jig lures wrapped in coils of copper line that were not cast but plopped. I fished at dusk, dawn, and night. I caught many beautiful cherry trout during this quest, but I never caught the Salmon. I even considered spearfishing and slaughtering the large fish out of spite. In the previous October, I had brought only my

Fig. 23. The karst topography of Samcheok County produced the "Big Hole," along with numerous caves and artesian springs that formed short-run trout streams. Photo by James Card.

mask and snorkel. I had found a tiny shallow spot near the edge of the Big Hole where I could kneel down, getting my legs, balls, belly, and chest wet, and peer into the Big Hole for a quick status check. Everything was normal, but some of the mature cherry trout had darkened bellies. The sides of cherries darken during the spawn, but this was far different. All trout in the world have white bellies, but these changed color. I never again observed the black-bellied trout.

I locked up the car and hiked past the water I usually fished on the way to the Big Hole and went directly to the casting spot, a small ledge on the river-right side of the pool. I scrambled over blow downs from typhoon squalls and made loose-hipped hops over flood-loosened rocks and bear crawled to the ledge. The closer you get to the water, the lower you have to be. No shadows thrown. No noise. Treat the fly rod like a rifle, the tip kept down and low. No false casts. Cast less, think more. I sidearm cast into the bubble line, and the caddisfly landed on a seam with a pert drift.

I watched the trout rise from the depths. There was a spurting take, and I had the fish on. I crawled backward off the ledge and eased into the water. I worked the trout out of the Big Hole and into the flatwater, where it could be netted. It was a lovely specimen, about a foot long, and I released it and it shot through the shallows and plunged back into the safe depths.

I broke down my rod and stowed it in my backpack. I had gotten my fix. I was so infatuated with the Big Hole that I never went farther upstream. It was such a fishing spot of great beauty and plentiful trout that it was hard to walk away from. But if a big hole was here, what if there was a bigger hole upstream? I scrambled over the wet cliff that surrounded the Big Hole and made it to the next level. I stood over the Big Hole looking downstream and looked at it with a new perspective. It was a jewel of a fishing spot, and there had to be more like it deeper in the valley.

The creek flowed over a limestone bed, offering pocket water, riffles, glides, and small drops. It was deep enough to hold trout but not big ones. What I was looking for was what I called "guide water," a stretch of trout stream that would provide at least a half day of fly fishing for a client. The Osip easily provided that, but the valley was in constant construction mode and, up and down the river, fishing spots were being blown out, so I sought untouched feeder creeks. I found them but they were not "guide water." Instead, they were tiny stretches of pristine slivers that held small populations of cherry trout that had most likely lived there since the last Ice Age. Those places were trout fishing perfection, but they were scaled too small. Many of these spots held only a short sequence of pools and riffles before the creek became too skinny to support trout life. Most of these sections were only a hundred yards long and offered only an hour or so of fishing but required more driving and hiking to reach, and it was not worth it to offer such an adventure to clients. On one unnamed creek there were only two crystal clear pools, two patches of pocket water, and three sets of riffles. A couple of casts into each spot and you were done. These were wild fish of secret spaces, of ice water and northern mountains, of lost valleys and water never looked upon. They embodied those species that live in remote and hidden hinterlands where the civilized do not go. Those trout are yeti. They are the Sasquatch.

As I hiked upstream the creek looked promising, and my step quickened as I scrambled over the ledges. Around a curve I found the source of the

Big Hole, and it stopped me in my tracks. Artesian spring water gushed out of a limestone rock face with the power of a loosened fire hydrant. I walked around the source of the spray and studied it from every angle. Water blasted from the rock as if it were a busted water main. I tried to imagine the underground channels that forced it to the surface at this particular point in this area's topography. I went past it and hiked farther upstream over an empty streambed. At one time water had flowed here but no more. I hiked farther and farther, and there was no creek. I walked back to the spring: that was it, game over. The creek that held the finest trout fishing spot in the country was only a short stretch that bulged with the Big Hole in the middle and a handful of fishable spots above and below it and no more.

12 The X-Trout of the Sohan Shut-In

In the morning darkness, I drove into the city of Samcheok and took a right onto the coastal highway heading south. There are beautiful beaches in this area, but they are blocked off by miles of chain-link fencing topped with concertina wire. Wedged into the chain links are stones to detect the disturbance of North Korean infiltrators. Spaced between the fencing are machine gun emplacements, watchtowers, and concrete bunkers. After the summer tourist season, the beaches shut down for the rest of the year and remain in a state that is both scenic and eerie. I pulled off on a village road that dead-ended at a beach. In the cold months, *myeontae* (Alaskan pollack) hangs from clotheslines and dries in the winter wind. During the late summer, old women spread out tarps on the macadam and dry rice and barley in the warm sun.

The vacant beach was my pre-trip and post-trip regrouping spot. As I readied my gear I liked to take a moment, stare out into the East Sea, and watch the sun rise over the waves coming in. Although I lived on the southern coast, I rarely got to appreciate a seafront that wasn't industrialized and wasn't infested with red tides that kept getting worse every year. Also the coastline was becoming infested with more and more giant concrete tetrapods, anti–beach erosion structures with four legs. They were another eyesore example of the construction cartel creatively using concrete as a solution looking for a problem. I laid out my gear, strung up my four-weight fly rod, tied on a black caddis dry fly, and dabbed on some Albolene. The first tentative raindrops of the day splattered every couple of seconds. I tasted the change in barometric pressure and felt the dropping temperature. It would be a cold rain. I slipped on my waders and a rain jacket and drove into the valley.

The Sohan Shut-In is a creek with few scars. It is amputated by a small dam and channelized near the village, but once the millrace ends, at the last farm shack, the stream is without blemishes. The upper section is one of the country's last remaining trout streams in a true roadless area, the last dream stream. I use the Ozark term "shut-in" to describe it because once you get back into the forested canyon and the narrow-as-a-hallway gorge, there is no turning back until you reach the source of the creek. Unlike all of the other rivers in the country, dying a slow death of a thousand cuts, the Sohan runs naturally for the time being. It's holy water because once something becomes rare it becomes sacred. The Sohan is a small, obscure masterpiece of how water sculpts the earth and how earth directs the flow of water.

The creek is a tiny blue scratch on a topographical map, and it is not difficult to find but very easily overlooked. The Sohan is a hybrid of a classic upland freestone stream and a spring creek fed by an ever-flowing artesian aquifer that provides clear water that is of constant temperature throughout the year. The gradient over its limestone and shale riverbed is nearly perfect, not too steep, not too lazy, and creating a riffle-to-pool ratio that provides hundreds of holding spots for trout. Tunnels of willows hold tightly to certain spots along the banks, and stands of bamboo cling thickly to the mountain slopes. There is one trail that leads back into the valley, to the Buddhist hermitage, and it veers up the mountain and away from the stream. Once on the stream you are completely alone, which is an incredible feat in a country that builds roads that parallel tightly to rivers.

I parked my car at the village and set off down the path, and I crept past a chained-up sleeping Jindo dog and prayed he wouldn't go into his usual spastic barking fit. The village was quiet and, when approaching one of my fishing spots, I preferred to slip in and out unnoticed. At the first pool I found a cylindrical fish trap sitting at the bottom, weighted with a rock and tethered by a nylon rope tied to a mulberry tree. I hoisted up the trap and opened the zipper. Inside were a dozen trout fry, ranging from one inch to three inches long, and a couple of stone loaches. I scooped their slick, squirming bodies and tossed them back into the water. Rotting carcasses

Fig. 24. The east-coast beaches of the Gangwon Province are fenced off to deter infiltration by North Korean amphibious commandoes. They are opened to the public briefly in the summer. Photo by James Card.

of fish that had wandered into the trap and starved were attracting others into the trap. The fish were swimming in through an inverted net cone and could not swim back out. Left unchecked or forgotten, the trap had become a perpetual killing machine that re-baited itself every time a fish swam in, starved, and died. I compressed the circular wire frame, folded it, and stowed it in my backpack to dispose of later at a city dumpster.

The trout of the Sohan Shut-In are cherry trout and rainbow trout. The rainbows of the Sohan are escapees from the fish hatchery in the village. The dam below the village blocks sea-run migration, and the trout of Sohan are landlocked, permanent residents of the shut-in. I slipped forward to a patchwork of shallow pocketwater that tends to hold only little trout. To grow any bigger they need at least a foot of water over them and a deep refuge within darting distance. I threw out the fly, and the mismatched crosscurrents grabbed the fly line and ripped the fly into the cataract that formed the pool below. I casted out again, with an aerial mend, upstream to give the fly a couple of seconds of natural drifting before getting sucked down again. A little rainbow popped the surface and took the fly and, upon feeling the hook, flipped twice into the air before I brought him to hand. I inspected the colors and markings of the fish, and I released him into the patch of water he had risen from.

I had an ulterior motive for coming here. The shut-in had taken the top spot as my favorite creek in the country, but now I had a mystery to solve. I was looking for a trout that was a mirage, a fish that might or might not exist. A year before, a handful of clients had a brief streak of catching some oddly colored trout on this creek. "Is this a cherry trout?" they asked. It had the distinctive parr marks of the cherry and coppery green back yet had the red band of a rainbow. The cherries have sparse but thick speckles that look like they were made by the thick tip of a black magic marker. A rainbow has tiny pinpoint speckles scattered over its back and sides. This trout seemed to be a mix of both.

"It's a rainbow," I replied and slipped it back into the water. "The coloration is a bit different in this creek." It happened more than once, and it left me feeling like a fool or, worse, a fraud. What kind of fishing guide cannot identify the trout he specializes in catching? So thereafter I decided to be honest about the fish and told them it might be a cross between a cherry trout and a rainbow trout, but I had no scientific evidence of the two species breeding in the wild. If it was a hybrid, it was an extremely rare trout. There are only a few mountain streams in the world that hold both rainbows and cherry trout.

This usually got the clients pretty worked up. Bastard game fish are highly regarded trophies among anglers. Tiger trout are a cross between a brown trout and a brook trout—fish of two different genus groups. They are loved for their pretty vermiculation—a wormlike pattern of squiggles across their bodies. A splake is a cross between a male brook trout and female lake trout. The cutbow is another hybrid prize, a cross between a rainbow and a cutthroat trout. The tiger muskie of the northwoods is a hybrid of a northern pike and a muskellunge, and they live up to their name with their colored stripes.

There were other puzzles to solve besides the unusual coloration and markings on the fish. Cherries spawn in the fall and rainbows in the spring, but spawning seasons can start earlier or extend later, depending on a myriad

Fig. 25. The X-trout, a possible *O. masou* x *O. mykiss* hybrid, discovered in Sohan Creek. Photo by James Card.

of environmental factors. And what of the offspring? Hybrids are usually sterile. Were the cherr-bows a result of a minuscule window of opportunity, a moment in time, when the spawning season of cherries and rainbows overlapped in the summer and the two fish became spawn-crossed lovers? And if their colorful offspring were sterile, they would live their short lifespan and die, never to exist again. They would have no legacy, and they would be unknown to the world unless someone documented their existence.

While researching the possibility of a hybrid, I came across a photo from a Japanese angler who claimed he caught an *Oncorhynchus masou* x *Oncorhynchus mykiss* cross. I sifted through hundreds of my photos of Sohan trout and found a couple of alleged hybrids, and they looked very much like the Japanese specimen. That was hopeful but inconclusive. I referred to the mystery fish in my field notes as the X-trout, using the mathematical symbol used to represent an unknown variable.

I contacted James Prosek, the artist and author of *Trout: An Illustrated History* and other books about the cherr-bows and sent him a couple of photos. We had corresponded a few times over the years about Asian trout.

James, in turn, contacted Dr. Robert Behnke, a professor emeritus at Colorado State University and one of the world's highest authorities on everything about trout. Dr. Behnke's opinion was that rainbow x masou hybrids were possible, but he did not see conclusive evidence that hybridization occurred. He wrote that, in his experience, natural intraspecific variation was a more likely explanation than hybridization.

My romantic notions of lovemaking among the rainbows and cherries were dashed by one of the world's greatest trout experts. "Natural intraspecific variation" would mean a naturally occurring difference that appears in members of a single species. I contacted Manu Esteve, a marine biologist from Barcelona based at the University of Toronto. He traveled the world studying the spawning behavior of salmonids by using underwater video cameras. I had made his acquaintance when Manu contacted me about the existence and whereabouts of Korean taimen in the Yalu River on the Chinese–North Korean border. He replied that hybridization between the two would not surprise him at all, but he doubted the offspring would be fertile. He pointed out that rainbows are spring spawners and masou are fall spawners, but the exact time of spawning can vary. He mentioned a similar case where he recorded small cutthroat male (spring) spawning with an adult coho female (fall).

The clouds darkened what little of the sky could be seen in the tight tree-canopied valley, and rain drizzled through the leaves. I fished the pools and deeper runs and skipped the pockets and shallow water. I eased up to another pool and climbed a small boulder and looked down. I spied five rainbows facing the current at the head of the pool. The more aggressive rainbows were always up front, edging out the cherries, which also liked the fast water. The relationship of cherry trout and rainbows in Sohan was like what happened when non-native brown trout were introduced in streams holding native brook trout. The brown trout tended to take over and dominate the brook trout. The same happened to the cherries, and they appeared in smaller numbers every year.

I pushed farther upstream, catching more rainbows than cherries. I spotted a rusty-coated Siberian weasel running along a moss-covered stone wall of the overgrown ruins of the old village—the foundations barely noticeable and almost completely reclaimed by the forest. The mulberry trees and the Japanese apricots the settlers planted long ago grew wild and unpruned. In the next two pools were two more cylinder traps, and I released the trout fry and stone loaches and packed away the traps. At another pool I found a set line tied to a bamboo pole wedged between two boulders. On the end of the line was a rotted six-inch rainbow. I snapped the bamboo pole over my knee and snipped off the line and hook. I scaled up the steep slope to a dead Mongolian oak knocked down by last summer's typhoon. I pruned off a large branch, fifteen feet long, with my hand saw and dragged it back to the pool. Out of my backpack came a length of kernmantle rope and a primitive rock-climbing chock made out of a big stainless-steel nut with a bike lock cable looped at the end. I waded into the water at the head of the pool, chest deep, and felt around under the whitewater spray. I found a crevice and inserted the chock and cinched it in tightly. Over the fat end of the tree branch, I lashed the rope and quickly pushed the branch underwater as hard as I could, using my body weight to hold it under. I knotted the cord to the underwater anchor loop and let go. The branch porpoised to the surface and straightened in the pounding current. It covered the upper half of the pool with half of its branches sticking out of the water.

The anchored branch would protect the trout from predators above, and the woody debris would attract aquatic insects that trout feed on. Anyone trying to throw a weighted hook or a bladed inline spinner into that part of the pool would get snagged. The only way to fish the pool without getting snagged would be to delicately place a dry fly along the branches and dance it through. Anyone trying to yank the branch out of the pool would end up confused and wet. They would be wasting their time. Wasted time means fewer fish would be poached. Nature protects the trout of Sohan in similar ways. In the summer the pools are protected by the webs of golden orb spiders. When gently casting, the spider web will snag a dry fly in midair, as if it were a natural insect. I have watched the Sohan trout fling themselves into the air to eat bugs caught in the spider's web.

I reached the Chute by midafternoon but, judging from the sliver of sky I could see through the forested canyon, it could have been dawn or dusk. Dark rain clouds formed overhead. The tallied results in my waterproof notebook were sobering but not as bad as I had thought. Poachers had cleaved the trout population, but there were still some wary survivors: many little ones that, if left alone, would grow big fast as the creek was so rich with insect life. There was a fishable number of what small-stream anglers would call average-sized trout—I caught ones between seven and ten inches, and I pulled in nine rainbows over a foot long and one lone beast of a cherry trout that pushed fifteen inches. Trout were still in there but small and not in the numbers they used to be. The glory days of the Sohan Shut-In were over, at least for a few years. Depending on the fertility of a stream, it takes about two years for a trout to grow three inches. By the time the fish is three, it should be around five to seven inches long. A trout taken on a small mountain stream that measures fourteen inches is considered a trophy. It might take up to six or seven years to achieve that size while surviving hooks and nets, ospreys, otters, bigger cannibal fish, floods, and droughts.

The dark-water pool before the Chute was a place I usually broke for lunch. I was a few hours behind my usual pace and famished. The rain continued to sprinkle, and my teeth chattered. There were micro leaks in my waders and rain jacket. On a warmer day this would not have been a concern. I removed a thermos from my backpack and gulped a cup of steaming *seolleong-tang*, an ox-bone broth. I guzzled down the rest and then knocked back a shot of sugared bourbon from an unbreakable plastic whiskey flask designed for boozing backpackers. The whiskey sting felt good on the tongue, and I took another sip before storing the flask. I was unable to peel the shell off the hard-boiled quail eggs I had brought to snack on, and I doubted my numb fingers could tie on another fly. I gnawed a mealy apple down to the core and flung it to the bushes.

It was getting late and I mulled over possible hypothermia, not at the moment but later. My entire body shivered. I needed to keep moving. I crossed the dark-water pool by edging along and feeling for the rough edges offering footholds. One misstep results in swimming. At the top of the pool, I pulled my body weight up on a branch and swung over the small waterfall and stepped into the Chute. One of my clients called the water of the Chute "fairy water, like something from a fairy tale." That day the sunbeams dappled the surface and penetrated to the pockmarked limestone streambed, and the water could put anyone gazing upon it into a trance. The Chute is about the width of a hallway. Its limestone walls are vertical until they reach the forest slopes of the shut-in. Inside the chute is a perfect glide that runs against the far wall, and the smaller trout at the end of the clear-water glide act as an alarm system. As soon as there is a disturbance in the chute, they bolt forward and put down the larger trout, at the head of the glide, that hide in a depression just under the water rushing down a staircase of gouged limestone.

I positioned myself to cast and watched the water. Raindrops perforated the surface, and it gave me a stealth advantage while giving the trout a concealment advantage. The largest cherry trout I ever caught in the Sohan Shut-In was taken from here, around sixteen inches. I flicked the black caddis fly on the glide and caught the drift. It was immediately snatched by a small cherry trout. I pulled him in and released him at the tail of the glide by submerging him and releasing him right at the opening of an underwater crevice. He went in there and stayed. I cast to the head of the Chute and let the fly bounce over the fast water and come back toward me. I spotted a shadow rising to the fly—a very nice trout—and dropping back down. I casted again and again but no looks. My hands were too numb to tie on a nymph for an underwater presentation. Seeing the shadow was enough, and I waded through the Chute, up the frothing knee-high staircase and into the most remote part of the Sohan Shut-In.

After the Chute is a series of wide, fast-moving step pools in a steep wooded gorge, where rocks peel off the mountainside at random and where cedar, pine, birch, and oak tend to die young if their roots don't purchase enough of the rocky earth to prevent gravity from pulling them downhill and into the stream. I had memorized some of the rocks and boulders along the banks, the loose ones, the solid ones, the slippery ones, and made good time upstream. I stopped to rest and cast out into a fast water run. The fly caught the right drift, and a bullet-headed rainbow smashed it within a second. I stepped back downstream and guided the trout into my net. It was not an X-trout. My theory was that the X-trout never made it past the Chute.

I needed to make time if I wanted to get out of the shut-in before dark. I bit off my fly, reeled in the line, and broke down the seven-piece rod and stuffed it in the rod case. I continued up the true right bank, and at the last stream crossing before the cave, I slumped down on the far bank and viewed it from a distance. Past the triangular opening, the cave is like a small cathedral. It's the source of Sohan Creek, and the spring water flows over the cave floor, covered with limestone roof-falls that have peeled off, dropped, and shattered. The spring water flows out of the cave, over a

Fig. 26. The headwaters of Sohan Creek flow out of the maw of a huge cave. Iron bars block the entrance, and cherry trout near the cave tend to be darker colored. Photo by James Card.

tumble of rocks, and into the very first pool of the stream, the Dogleg Pool. The current flows straight out, collides against the rock wall, and turns to flow down the valley.

It's the end of the road unless you are into cave diving. But for that, government permission would be needed. The cave was sealed off by a grid of steel bars bolted into the rock. There was something haunted about the mouth of the cave, and above a small door with a padlock hung a sign that prohibited entrance, warning punishment: a twenty million won fine or two years in prison. Bats fluttered in and out, and at one time some dark-colored cherry trout lived near the entrance, but they haven't been there in a few years. I climbed the steel bars once to the top of the cave and shined a flashlight into its depths. The floor of the cave was about shin-deep, and a dark hole oozed out thousands of gallons of cold, clear spring water. I once had a dream that Japanese colonialists had stashed a hoard of stolen gold in the wellspring.

Above the main cave is a smaller cave about fifty feet up in the jagged cliffs. I tried to access it once and almost plunged to my death and vowed never to try again without rappelling gear. In front of the Sohan Cave I once found the bones and hair of a musk deer or water deer—they are similar in size and shape. I have never seen a deer or its tracks past the Chute. There was no explanation of how its remains ended up in front of the cave unless it fell from the cliff.

13 The War against Rivers

Pyongyang's state-run Korean Central News Agency was saber rattling again, threatening to turn South Korea to ashes and a sea of fire. They had recently test-fired some Rodong missiles that landed harmlessly in the East Sea, but they had gone further than ever before and that made the military pundits and diplomats very nervous. The younger generation mostly ignores this, but those who remember the war still head to the supermarket and stock up on food.

Among expats, Seoul was nicknamed "Killbox 90201," with the sitcom zip code added as a joke. The surrounding urban mass of the Gyeonggi Province was called the "scud belt." If North Korea were to attack, an estimated one-third of Seoul would be leveled within an hour, and one million people would be dead the first day. That's just the artillery and tube rocket launchers. That's not counting chemical warfare, the ensuing firestorms as Seoul burns, or havoc created by fifth-column saboteurs.

I heard of U.S. military plans for evacuating American civilians, but I never had much faith in it. We were supposed to listen to AFKN radio for instructions on where to go, but the pecking order did not look good for me. First would be military dependents, civilian contractors, Department of Defense officials, embassy staffers and family, ill and disabled people, and then regular American citizens. That is also not counting executives of multinational corporations and their families and wealthy South Koreans who bribe or influence their way into the evacuation lines. I kept an emergency bag tucked in the bedroom closet. It was a medium-sized backpack with birth certificates, diplomas, marriage documents, passports, bank account documentation, a small hard drive, two tanto knives, socks, work gloves, a spare Leatherman, a water purifier, a flashlight, rolls of South Korean and American cash, two toiletry kits, and a week's worth of MREs I had purchased from the black market. There are black markets in South Korea that look as if an entire American grocery store had been looted and rearranged in a shambolic mess composed of tent-like mazes of small shops. Everything there had been stolen off U.S. military bases. The South Koreans who are employed on U.S. bases in the Army and Air Force Exchange Service are the largest organized criminal group in the country. It's a multimillion-dollar industry, and it's been going on since the end of the Korean War. Every base in the country is regularly ripped off, and the U.S. taxpayer foots the bill.

If Korean War II broke out, I planned to live on my wife's home island, look after her family, and venture out to report on the war as a freelance foreign correspondent. I figured I would do very well as my competition did not know the Korean backcountry as well as I did. There would be parachute journalists coming in from other countries, but they would be clueless and need hand-holding. Many of the in-country international media were members of what I called the Hibiscus Club. It was made up of journalists, pundits, and scholars. They were often apologists for South Korean interests in return for status, access, or career building. They all resided in Seoul and were very soft and fragile since they did not venture into the field.

Sometimes I had to go to Seoul, and this trip was to check out a small man-made stream and to do some urban fishing. I stopped in Itaewon for a cheeseburger. It was the only neighborhood in the country that served decent Western food. Itaewon is Planet Earth's version of Mos Eisely. There are bananas, Asian gangsta wannabes, yellow-fever yard apes, sidewalk pissers, Philopon dealers, K-pop bimbos, deranged whoremongers, fresh-off-the-boat Koreaboos, look-twice ladyboys, big-nose roundeyes, paleface Yankees, Gangnam-style pimps, dog-faced grunts, Fulbright scholar parasites, street photographer simps, limey wankers, hymen restoration stitchers, massage-parlor indentured servants, Nigerian princes, grip-joint barbershop *ajjumas*, fanny-pack tourists, plastic surgery victims, vagrant ESL teachers, counterfeit-purse hustlers, sidewalk pukers, wedding dress–wearing hookers, Lonely Planet beaten-path followers, soju dipsomaniacs,

Chongnyangni 588 rejects, birthright-citizenship scammers, Filipina torch singers, mail-order Natashas, degenerate diplomats, rice rocket poseurs, vagabond backpacker narcissists, blacked-out Korean salarymen everywhere, flocks of pigeons eating last night's vomit, and a handful of normal people.

I loved to visit Itaewon, but I took it in small doses. My querencia was in the deep mountain backcountry, but it was nice to check the pulse of the nation's capital now and then. After wolfing down the cheeseburger, I grabbed a taxi to visit the Cheongyecheon, the most famous small stream in the country. It was the pet project of Seoul mayor Lee Myung-bak, a former Hyundai Construction CEO nicknamed the "Bulldozer." He "daylighted" the stream, the term for opening up a waterway that was once buried by urban development. After the Korean War, it was a shantytown sewer ditch, and it was paved over and then the Cheonggye Expressway was built on top of that. Lee pushed to restore the stream, and when it was finished, the project was lauded in the media as a great achievement in urban renewal and environmentalism and a creation of a cultural and recreational space.

The artificial stream was 6.8 miles long. I hiked half of it and had seen enough of the fake creek. It is an ephemeral stream, flowing only during the rainy season, so water is pumped in to maintain a constant flow through-out the year. It is another concrete canyon, and I supposed that was good enough for the burghers of Seoul. I spotted a small common carp, and some nature-deficit-disorder school kids squatted and stared at some minnows. With landscaped greenery, small bridges, stepping stones, and footpaths, it wasn't too bad a place to go for a stroll, but it hadn't been restored to any kind of natural state. It was a landscaping project like a cross between a cascading backyard water fountain and a man-made lazy river for tubing at a water park, but on a metropolitan scale.

There was something more sinister afoot. Restoring the Cheongyecheon was a setup, a compliance test, a trial balloon to gauge public perception on manipulating a river system under the guise of phony restoration work. South Koreans have forgotten what natural rivers are supposed to look like, so their culturally induced ignorance (landscape amnesia) was exploited.

In a puff piece written by Bryan Walsh, TIME magazine named Lee a "Hero of the Environment." Lee Myung-bak was now South Korea's pres-ident, and he parlayed his newly minted eco-friendly reputation to pitch the Pan Korea Grand Waterway to the public. It was a vision of a giant canal system spanning the country, and the Han, Nakdong, Guem, and Youngsan Rivers would be dredged, channelized, and fitted with locks and dams for cargo and freighters, and one section would have a thirteen-mile tunnel going through a mountain. It expanded, and seventeen other small canals were added to the project. Part of the project would be funded by selling sand and rock dredged from the riverbeds to a domestic market already well supplied with aggregate material. It reminded me of a Saturday Night Live skit where John Malkovich plays Johnny Canal, a lunatic coonskin cap–wearing frontiersman proposing to connect all cities, towns, and villages in America by thousands of individual canals. When questioned about the absurdity of the scheme, he goes apeshit with a hunting knife.

Shipping and logistics experts thought it was unnecessary. Surrounded by water on three sides, South Korea has plenty of ports, and the interior is a network of railways and roads for efficient transport. Canals are used to transport bulk materials, and nobody was clear on what materials would be shipped or for what reason. Having grown up along the Mississippi River, I considered the draft and how low water affects the navigability of barges. There were many shallow stretches of the Nakdong where even fishing boats would be scraping bottom.

It was opposed by environmentalists, academics, Buddhists, and Catholics, and the South Korean people did not fall for this boondoggle, except those that sought to profit from it through riverside land speculation. Lee quickly repackaged the scheme as the "Four Rivers Restoration Project," and instead of being turned into a canal the Han, Nakdong, Guem and Youngsan would be "restored." The usual tropes were trotted out in the media: improve water quality, prevent water shortages and floods, and eco-friendly this and that. It was called the "Green New Deal." It would provide many hundreds of thousands of jobs, and supposedly many foreign tourists would come and visit these eco-friendly cultural green spaces. It was also mentioned without explanation that this project would somehow fight climate change.

More dams would be built, sixteen sluice gates would be installed, old dams would be rebuilt, the thalwegs would be dredged and leveled, "super levees" would be erected, and the riverbanks would be armored. As this garnered controversy in the public space, fifty smaller rivers were earmarked for the same treatment, including the Seomjin River, and finally the Land

Ministry announced they were making a plan to revitalize all rivers in the country. They would also pave 1,073 miles of bike trails along the rivers. Repackaging the canal project as a fake river restoration project made it obvious that this whole enterprise was not about creating a canal system for cargo transport or anything about true river restoration. It was about raiding the public treasury for government contracts to wage a war upon nature.

The Four Rivers Project was officially started in late December of 2008 at a riverside ground-breaking ceremony in the city of Andong. The Nakdong was the first of the four rivers to get the shovel. It had been underway well before then. In the past months I had been seeing excavators, bulldozers, and dump trucks digging and leveling smaller rivers throughout the country. I do not know whether they were part of the Four Rivers Project, but many smaller rivers were added to the project's mission creep. You travel and explore the Korean river valleys enough and a backhoe digging along a river is a common sight, but the heavy machinery all appeared very quickly, in a well-orchestrated fashion, and in greater numbers than ever before. I followed the Moscow Rules: Once is an accident, twice is a coincidence, and three times is an enemy action.

I visited a neighborhood in Seoul that I had never ventured into. It was one of those subway stops that a person passes by his entire life and never gets off. There's no reason to. But now there was someplace I needed to go. I limped through the streets. It was the standard mix of Seoul's antlike middle and lower-class living—dingy mom-and-pop grocery stores, eateries, soju dives, and hundreds of midrise Brutalist apartment buildings that housed way too many people in way too small a space. It's a part of Seoul that ages people prematurely. Side streets and alleys fingered through these warrens, and I zeroed in on the address I was looking for. My lower back was flaring up, and I was delirious with the pain. The bouncy and rocking subway ride had made it worse, and climbing the subway stairs to the Earth's surface had been a slow struggle.

While living on Koje Island, I had hurt my lower back while lifting a heavy tub. The pain was nothing like any kind of physical injury that I had incurred before. It put me on my knees, hunched like a wounded animal with an arrow stuck in its back. I ended up walking with a cane. One of my students told me to go see Dr. Kim at the Daewoo shipyard clinic. When I

Fig. 27. High profile construction projects on South Korea's major river systems get media attention, but the destruction of thousands of small rivers throughout the countryside do not. Photo by James Card.

arrived, Dr. Kim took one look at me and smiled and nodded. He set aside the cane and asked me to touch my toes—an unbelievable request. I tried it anyway, and then he had me stretch in other positions.

He didn't speak much English but he said, "ky-ro-prak," and he led me to a massage table where he torqued my body this way and that way. He helped me off that table and led me to another one. He pulled out a tray of needles. I had never had acupuncture before. I didn't care for needles, and I always accepted them with gritted teeth. He placed the needles in my lower back. They were wispy thin, and he gestured for me to roll over. I protested, thinking the needles would be stabbed deeper into my back. He said, "It okay," three times in a row and placed a heating pad under my lower back and affixed some electrodes to it. The needles flexed with me lying on top of them. The pad was wired to a small control box, and Dr. Kim cranked up a dial. The needles shot electrical pulses, and the pad warmed up. I entered a state of electrified delirium, like an LSD trip for a backache.

Around twenty minutes later, Dr. Kim came back and unplugged the machine, pulled out the needles, and tossed them into a metal tray. He told me to stand up. Stand up I did. He told me to touch my toes, and I did so with no pain. He smiled like a magician who had pulled off an old magic trick before a new audience. I walked out of the clinic at 100 percent physicality. I felt like doing deadlifts and splitting firewood. Dr. Kim did this without x-rays, nervous system scans, or medical questionnaires. He did not check my pulse or blood pressure or height or body weight. He was the best doctor I had ever encountered in my life.

After that, my backache would flare up about once a year. I visited other acupuncturists, but none of them were as good as Dr. Kim—but since I no longer worked at the shipyard, I could not visit him. One quack performed moxibustion with glass bell jars and burning rice hulls and left perfectly round welts all over my back. It looked like someone gave me hickies with a vacuum cleaner.

I was getting closer to where I needed to be in this new and obscure neighborhood. I stopped at a mom-and-pop store and bought three bottles of OB Lager. The alcohol would help with the back pain, and I had to make this work. I had to make this fishing trip a success no matter how much pain I endured. I was on a roll as a freelance journalist. Magazine assignments were coming in, and I'd just had two big wins: my first story in the Asian edition of the *Wall Street Journal* and my first story in the *New York Times*.

The story for the *Wall Street Journal* was a travel piece about an American man named Carl Ferris Miller who came to South Korea long ago, decided to stay, and settled on 158 acres of seaside land on the Taean peninsula. Over the course of his life (he died, in 2002, at the age of eighty-one), he turned his land into a world-class arboretum.

Earlier in the year I had visited Singapore and fished for peacock bass, a predatory cichlid that exists in the jungles of South America. Other than its native waters and a few other places, such as south Florida, where it was introduced, it is hard to find. When I learned that it was stocked in the urban reservoirs of Singapore—of all places—I knew I had to go after them.

I decided to write about this experience "on spec," meaning you research and write the story on your own without an assignment and, after it is done, shop it around for takers. If a publisher buys the story, perfect. If not, you just wasted your time and energy writing a story nobody ever asked for.

In the piece, I juxtaposed the irony of sober and stuffy Singapore stocking their ponds with a foreign take-no-prisoners apex predator game fish. To be a dick, I sent it to the *Far East Economic Review* because the magazine was banned in Singapore.

I also sent it to *FEER* because it was a publication I had long admired. They were known for their Asian business and political coverage, but they sometimes published travel dispatches from around the Pacific Rim. As I waited for a reply and worked on other projects, I learned they had closed shop after sixty-three years of publication. The owner, Dow Jones, laid off the staff, and my story presumably was in an email inbox that was now a digital crypt.

The best advice I had picked up along the way as a freelance journalist was: always be pitching stories to publications that you have no business pitching stories to. I found the name of an editor at the sports desk at the *New York Times* and fired off the Singapore peacock bass story. I had been rejected so many times as a writer that punching above my weight class was now instinct.

Meanwhile, I was an avid reader of the *International Herald Tribune*. It was first published in 1967 and owned jointly by Whitney Communications, the *Washington Post*, and the *New York Times*. In 2003 the *Times* became the sole owner of the paper. Stories that were published in the *New York Times* were sometimes republished in the *International Herald Tribune*. I kept noticing the byline of Michael Brick. He was a brilliant writer and a solid reporter, and he tackled the kind of stories that I was interested in. They appeared in a column called "Pushing the Limit": rock climbing, off-road racing, rattlesnake hunting, duck hunting, snowmobile racing, powerboat racing, bowhunting elk, shooting doves, mountaineering, and winter survival runs—to name a few. Who was this guy?

There was also one other thing at the *Times* that caught my interest. Nelson Bryant had retired just a few years before, in 2005. He had been writing the "Outdoors" column since 1967, and now there was a void. I knew that newspapers across the country had been quietly killing off their outdoor coverage. Bryant had never been replaced, but it seemed that the *Times* was still committed to running some outdoor pieces now and then—mostly patchwork produced by staffers and freelancers. If that beat was vacant, then I wanted to move in and claim some of it as mine.

I received a reply from one of the sports editors. He wanted to run the Singapore story. The story ran as "Fishing in Singapore for an Anti-Singapore Fish," and my favorite lines were: "The government may dictate natural order by landscaping the city-state as a well-tended but very unwild suburban park, but it cannot control the underwater jungle of the reservoir ecosystem. There, the peacock bass is the predator king."

With the Singapore story published, I wanted to strike while the ink on my byline was still wet, before the busy editors forgot my name. The competition was fierce to get your story published in the Gray Lady. The competition was not just other writers, it was other sports stories: celebrity scandals, world championships, up-and-coming pro-sport underdogs, and I could not compete with that.

The only way I could compete was to work the fringe and come back with something that nobody had ever heard of before. It was the only thing I had in my pocket to bet with: original reporting from the field—from places no other journalists were sniffing around. I wanted to prove that my success wasn't a fluke, that my byline was worthy of a second shot. I didn't want to be a one-off. The follow-up had to be good, and it had to stand out. I fell back on my strengths and Twain's dictum of writing about what you know: I pitched the most batshit-crazy fishing story I could think of, and I got the assignment.

I was closer to the address, and I could see the sign to the place: Gold Indoor Fishing Spot. I let the knot in my lower spine become numb, and I swallowed the pain and let it dissipate throughout my body. I stepped down the basement steps and entered. The room was dimly lit and had dark walls and a black ceiling. In the center of the room was a pool with walls just under waist height. The water was murky and aerated, and it had some circulation. Surrounding the pool was a countertop with an ashtray, and seats were spaced around it. I rented a station for an hour at a cost of $8. The attendant gave me a small fiberglass rod tipped with monofilament and rigged with a glow-in-the-dark bobber, a sinker, and a small hook. I was also provided a net and a towel. The bait was a bowl of fish kibbles.

These places had been on my radar for a while, but I had never ventured into one. I had seen them throughout the country. There was one near Changwon that was covered with a translucent vinyl tarp used for greenhouses. Anglers sat in lawn chairs next to the water's edge as if fishing in

Fig. 28. Anglers fish for carp at a basement pool in a Seoul neighborhood. Photo by James Card.

a swimming pool. Two other anglers were fishing, and occasionally they would get a strike. One angler missed three strikes in a row and would mutter, *gaesaekki* (son of a bitch), but he always said it with an I-will-get-even smile. The pool was stocked with crucian carp, Israeli carp, common carp, catfish, goldfish, and koi carp.

I caught a few fish, and I enjoyed it in a simpleton way. I also participated in a game where the fish was weighed on a scale and the results were put up on an electronic scoreboard. I won a set of two ceramic rice bowls. After I got enough notes and photos for the story, I hiked back to the *yeogwon*, where I dumped my overnight bag. I had bought more beer along the way, and I was feeling better due to the alcohol but also because I had gotten my work done. From a street vendor I picked up some *soondae* for a late-night meal. The linens on the bed at the *yeogwon* were clean and crisp, and lying down took the pressure off my lower back.

In the morning my back felt better after some stretching. Sitting made the back pain worse, so I picked up the bedside table and placed it atop the small coffee table, making a stand-up desk. I pulled out my notes and started to write. If all of the river systems of the country were going to be ravaged by a misguided construction project, the fisheries would suffer and so would the angling. I was reporting from the present, but this was a glimpse of angling in a dystopian future.

14 The Child, the Dog, the Ape, and the Dead

The day my son was born, I was up early to fly fish for snakeheads. As I was walking out the door, my wife told me her water had broken. I knew this was important, and I knew that I would not be fishing. In the hospital waiting area, I paced around and scratched my hands and forearms. An odd rash had emerged overnight and was getting worse. I pawed the suppurating pustules like a leprous geek and connected the dots. The evening before, my wife had asked me to prune the neighbor's lacquer tree that was growing over our courtyard wall. She said it was poisonous and she was highly sensitive to it. I'd hiked through groves of the exact same tree many times before in the mountains with no ill effect. As I pruned the trespassing limbs, a sap oozed from the snipped branches. I gathered them up and tossed them in a compost pile.

The lacquer tree (*Toxicodendron vernicifluum*) is the Asian cousin of poison sumac. It's in a genus of trees, vines, and shrubs whose most infamous member is poison ivy. These plants produce urushiol, a caustic oil that severely irritates the skin. The thing about urushiol is that the oil can be passed by indirect contact such as touching the blades of pruning shears, petting a dog that just went through poison ivy, or even cloth-to-skin touching. I'd been immune to urushiol my entire life, but this exposure was too much for my system to bear.

A nurse called me in, and I went to see my wife. She was in labor but another thing was making it worse. Earlier, while comforting my pregnant wife, I had touched her legs and arms and transferred the urushiol to her skin. She now had small patches of rashes. She hissed at me not to touch her and called me a monster. When my son was born, hours later, I was ordered by the doctor, the nurses, and my wife not to touch the infant.

I drove her to her childhood home on Koje Island, where her mother would look after her and the baby while she convalesced. There is a post-partum tradition that a new mother should get complete rest and bond with the baby. This can last a month or two. She would eat huge amounts of *miyeok-guk*, a seaweed soup that is a restorative. She would not leave the house and would stay warm, avoid cold liquids, wear socks and warm layers, and cover herself with blankets and hot pads.

It was useless for me to stick around, and my touch was a transdermal liability. I drove back to Changwon for a last period of bachelorhood. It was just me and the dog. When my wife had hinted at wanting a child, she had said taking care of a dog is a great way for a young couple to get used to the responsibility of parenthood. I agreed but on one condition: it had to be an outdoor dog, and it would be my fly-fishing companion in the backcountry.

A chance encounter at my neighborhood mom-and-pop store led me to him. A man had a cocker spaniel on a leash, and her teats were almost dragging on the sidewalk. I asked if she had just thrown a litter. He looked like a Korean version of Joe Pesci playing David Ferrie in Oliver Stone's film *JFK*. He gave me his address.

I stopped by later that afternoon and looked at the only male in the litter. His littermate sisters clambered over him and pushed toward me. I whistled and clucked my tongue, and he perked up, curious. I removed him from the other pups and held him. He squirmed at first but relaxed at the sound of my softened voice and looked into my eyes. I studied his muscles and silky coat and pried open his mouth full of sharp puppy teeth, and all were fine. I removed a pheasant wing feather that I had found in the woods the day before. It might have still held scent, and I brushed it under his nose. He came alive. I set him down and waved the feather before him. He rushed it. I dangled the feather above him and let go, and he pounced on it as it touched the floor. He was the one.

I named this unpapered American cocker spaniel puppy Blitz. We let him get used to our new home for a week, and then I started training him. Daily walks both on leash and heeling, simple retrieving drills, and learning voice, whistle, and hand signal commands. While hiking along a trail in the mountain foothills, he flushed his first pheasant. He was nine weeks old. That moment changed our lives.

Much like the untapped fisheries that I discovered, ring-necked pheasants (*Phasianus colchicus*) in South Korea are as bountiful as the legendary pheasant populations of the Dakotas. They were everywhere. I knew this as I accidently flushed them now and then when fishing. It wasn't until I had Blitz to flush them for me that I realized South Korea had the potential to be a world-class pheasant hunting destination. This is because of the superabundant numbers of birds, but also the extreme terrain is unlike what most people attach to pheasant hunting. In America, it is an upland sport of walking miles over flat fields and endless prairies. In South Korea, it is closer to chukar hunting, and going after pheasants is a game of climbing steep sidehills and brush-choked mountain slopes.

For Blitz, the mountain foothills became a place where the moments spent afield were measured by "making game." Finding and flushing upland game birds is what spaniels are bred to do, it is their truest calling, and letting Blitz do what he was genetically hardwired for allowed him to achieve a state of self-actualized grace at a very young age: he lived for chasing ring-necked pheasants. After his first one hundred or so flushes on pheasants, he was a muscular, barrel-chested puppy. His puppy appearance still clung to him, especially lounging about the house on the *ondol* underfloor heating. He knew of the cool spots, the warm spots, and the hot spots, but when he took to the woods, he became an invigorated hunter.

By the time Blitz was one year old, he had flushed over a thousand wild pheasants, the result of daily hikes, in all weather—barring typhoons and monsoon rains. Sometimes twice a day, often the same pheasants were flushed, from the same family, on the same mountain, on the same slope. For Blitz, pheasant scent was olfactory ambrosia, like snorting wildland cocaine. Along with pheasants he flushed coveys of Coturnix quail and hazel grouse while I was fishing in the Taebaek and Samcheok regions.

During this quiet time of man and dog alone in the house, after my son was born, I decided to take care of some unfinished business. There was one species of native Korean fish that partly eluded me. It was the northern snakehead (*Channa argus*, in Korean: *gamulchi*). I had caught a few before when bass fishing and I was impressed by this vicious predator, but I had never caught them in great numbers and always by accident. I wanted to purposefully target them rather than hook them as a bycatch.

I got a tip that a farmer was raising them in a commercial fish farm near the Junam Reservoir outside of Changwon. Junam and the adjoining Dongpan and Sannam Reservoirs were among my local go-to spots for giant bluegill. These bodies of water had provided for many Friday night fish fries. I guided a few trips for American sailors from the Chinhae Naval Base through their Morale, Welfare, and Recreation program. I got them suited up in waders, walked them through a quick fly-casting lesson, and let them get after invasive largemouth bass and bluegill. I never specifically targeted snakeheads in these waters, although I knew they were most likely in there and if a snakehead fish farm was nearby, there was a chance that some of them might have escaped and bolstered the wild population.

If there were any kind of fish that could pull off a jailbreak it would be the snakehead. Its respiratory system allows it to exist outside of water for days, and it can slither overland to another body of water. I have never seen the fish wriggling on land, but once I found its belly tracks on a muddy beach and there was no other explanation of what creature could have left those marks. The snakehead has an American doppelgänger, the bowfin (*Amia calva*). They are similar in life cycle and morphology, but the snakehead has better-camouflaged scales. The bowfin is an ancient fish, a living fossil, and fly anglers in the know have the greatest respect for them as fighters. They may be the most underrated freshwater game fish in North America.

The reservoirs of the Junam area are the nation's biggest inland stopover for migratory birds, and environmentalists have recognized the area as a wetland of international importance although they are nothing more than nondescript swampy man-made lakes. In the 1980s the military government drained most of the inland marshes, dammed rivers flowing into the sea, and destroyed the brackish estuarine zones of tidal flats and coastal wetlands for agriculture and industry. This was a place to be preserved but not lauded as an eco-destination, as it is in the Korean press. It's the last of a once-great habitat, a lone waypoint at the far end of the peninsula before the ocean. The migratory birds arrive here because there is no other place to go.

Last winter, in the surrounding hinterlands covered with hoarfrost, I let Blitz flush his fill of ring-necked roosters and Coturnix quail, then we moved closer to the reservoirs. I leashed him up and had him heel as I bird-watched

Fig. 29. The northern snakehead, *Channa argus*, a fearsome predator fish yet kept very much in check by largemouth bass. Photo by James Card.

with my binoculars. Skeins of white-fronted geese and taiga bean geese glided into the rice stubble. I saw Eurasian widgeons, Baikal teal, spot-billed ducks, gadwalls, mallards, pintails, northern shovelers, moorhens, scaup, coot, bitterns, and swan geese. The highlight was in March, on a day with bone-chilling fog. White-naped cranes arrived from Manchuria, and they stalked through the cold gloam like wraiths in a cemetery.

Blitz tracked me along the shoreline as I slunk into the primordial water. This was a far different angling experience than fly fishing for trout in clean cold-water streams. The lake bottom was composed of hard sand ledges but also pockets of muck that released a swamp gas odor when muddled up. It was an aquatic jungle, and it made me a true believer in the aquatic ape hypothesis—the idea that our early human ancestors shifted their existence to live close to water. Academics debate it but from a wilderness survival perspective, everything was here: freshwater, mussels, clams, turtles, frogs, fish, waterfowl, and edible greens. The aquatic apes were hunter-gathers that speared fish in estuaries and raided duck nests in wetlands. They dug clams in tidal flats, hunted game trails that paralleled rivers, trapped trout in creeks, and harvested wild rice from lakes. Water was life, and I spent so much time in it that I often felt an ancient aquatic ape-like connection.

I eased waist deep into coontail, floating water moss, water caltrop, native frogbit, yellow floating heart, sacred lotus, heartleaf false pickerelweed, water hyacinth, bulrush, arrowhead sagittaria, blue iris, and Japanese bindweed. It took me two summers of fishing to identify those plants, bringing home specimens one by one. There was one plant of special interest: the prickly water lily. The round leaves were three feet across, and the surface had a rough texture like the back of a scaly green lizard. They were floating mats that acted like protective roofs, and predator fish loved to hang underneath them, much like how game fish are found under boat docks.

I was solely targeting snakeheads, and on my six-weight I knotted on a froggy deer-hair humper. I cast it onto a water lily mat and, upon landing, the gentle impact sent a wave ring around the floating mat. I let it sit for a long minute then twitched the fly to the edge of the mat and let it hang for a moment. I plopped the fly into the water and raked it across the surface with a fast strip. There was a burst of water, and I set the hook: a largemouth bass. The fish tail-walked across the water trying to throw the hook. I kept the rod high and the line tight and brought the bass to net. I looked over at Blitz. The splashing had his attention, and he was at the waterline. It was a nice twenty-inch bass, fat-bellied and healthy. I released the fish and wished it were a snakehead.

The next two hours of casting produced more and more bass until I was sick of catching them. Blitz milled around on the hillside shore and followed my movements. I had a little time left before I needed to get back home, so I swapped leaders and tied a small woolly bugger to the tippet. I waded over to the Mariana Trench, a spot that I had nicknamed because of a long open-water gap between mats of coontail where bluegill lurked. The water was always deep and dark, and there were some spots where I had to stand on my tiptoes. The first cast brought in a slab-side bluegill that I could barely get my hands around. I folded down the fins, gripped the thick-bodied fish, and plucked out the hook; then I slid the fish into my old spearfishing net bag that was tied to my waist. The fish's zigzagging and sharp dives stirred up the water and seemed to liven up more bluegill. A feeding frenzy was on, and the next cast brought in another. I collected nine keepers for dinner and hiked back up to the embankment road.

The next three weeks I fished Sannam and caught hundreds of bass and had two strikes that were unmistakably snakeheads. The explosion of water was unlike a largemouth bass, even a huge one. Each strike was a lunge or a charge that sprayed water forward. Both times I did not get a good hook set. The strikes were so violent and immediate they made me flinch. I quit fishing the floating mats of water lily—the snakeheads simply did not lurk there—and I focused on the weediest tangles I could find. Each of the two strikes was as if a semi-submerged log had come alive and blasted forth like a torpedo. I modified my flies with a small plastic weed guard to keep them from getting snagged in the mess, and then I finally caught one.

This strike was the same as the other two, and the fish fought violently and in a way that reminded me of a northern pike. Getting it close, I did not bother with the net and grabbed it behind the gill plates. I walked it to shore for some photos. It was meaty, beady-eyed, snakey, toothy, and so mean and ugly looking that it garnered my respect as a fighter but also as an evolutionary survivor, and it had a natural beauty that Darwin would love. I released the snakehead into a patch of frogbit, and with one tail flick it disappeared.

Here you can see Gause's Law in action. The theory states that when environmental factors are constant, two species competing for the same resources (food, habitat) can't stably coexist. One will have an advantage over another. That will lead to the decline of the less-competitive species, or it will be forced to adapt in some way. This is occurring in real time with largemouth bass and snakeheads. I wanted to catch more snakeheads but without the interference of largemouth bass. I had to find a swampy lake uninfested with bass that held a stable population of native snakeheads.

The hand wringing and hyping of lazy science journalists writing about snakeheads taking over North American aquatic ecosystems are overblown. As long as they share the water with native largemouth bass, the bass will keep their reproduction in check at near-decimation levels. The ignorance and anxiousness of American scientists that have studied snakeheads is easy to explain: they have viewed this creature only while standing upon American shores. If they flew over to South Korea, they would find bodies of water where largemouth bass, bluegill, and other Korean native species have coexisted for decades. The bass always becomes the main predator in the food chain with only a few snakeheads growing to a sexually mature size until the voracious bass is no longer a threat but, perhaps, lunch. Snake-heads should be celebrated. Brown trout and ring-necked pheasants are non-native species beloved by anglers and hunters. The snakehead should receive the same sporting admiration.

It wasn't long after my wife and son came home that I found the dead man. My wife complained of a bad smell emanating from behind our house, and she accused me of urinating on a pile of Zen garden landscape rocks. That was true. I pissed there every day, but it was not the cause of the smell. It was the scent of decomposing flesh. My theory was a cat got hit by a car and scrambled into our courtyard and holed up somewhere to die. I did not find the cat, but I noticed blowflies buzzing around a screen window of the house that abuts our courtyard wall. I cupped my hands and peered in and discovered him.

He was sprawled on the entryway steps inside his tiny apartment. His legs were blackened with rot, and maggots dripped off him into a pool of putrefying body fluids. I stepped back from the window and dry heaved and looked back through the window to double check and dry heaved some more.

The police worked fast, and within two hours he was gone. The ashamed landlord had no idea that his tenant had died and had lain there rotting for so long and that nobody else had looked for him. He died alone. There was speculation he might have died of fan death, a belief that sleeping with an electric fan operating in a closed room will kill you. I never learned the man's name or how he died. He was in his sixties and had moved into the apartment, which had no air conditioning, in the pounding August heat, only two weeks before. I didn't have a chance to get to know him. Spooked about the incident, my wife asked the monks at the local Buddhist temple to offer prayers to help his spirit go to the place it needed to be.

I wondered what happened to the dead man because Blitz and I spent a huge amount of time among the dead. Koreans bury their dead in the hills, and scattered on the mountainsides are millions of earthen mounds in small clearings in the forest. It's here where we found and flushed most of the ring-necked pheasants.

Imagine if all the graves in all the cemeteries in America were dispersed in the woods, hills, and mountains throughout the nation. And on every hike, on every hunt, on every fishing or camping or bird-watching or backpacking trip you come across graves. They become part of the outdoor experience. The dead are not shunted into the confines of a graveyard but exist among the living and are part of the landscape. When I am out hiking, some graves become mnemonic waypoints, and some are hard to see because they are overgrown with multiflora rose, raspberry thickets, and lespedeza—which, in turn, attract wildlife.

When a person dies and is to be buried, a small meadow is cleared on the forested slope, and the corpse is placed within an earthen mound. Percival Lowell described them as "Huge porpoise-backed mounds," in *Choson, Land of the Morning Calm: A Sketch of Korea in 1886*. Some families ring the humpbacked mound with granite or marble blocks and front the mound with a polished black marble tombstone engraved with Chinese characters. Sometimes you see a few with a guardian statue on each side of the grave, which become green with moss. Some graves are completely unmarked, just a grass-covered mound, presumably from a family of humble means. Royal tombs are sometimes protected by statues of *haetae*, a grotesque scaly leonine beast that resembles a hodag.

Confucian custom dictates that the men of Korean families are required to maintain the mountain graves of their ancestors. Some families are not so dutiful. They come apart through divorce, death, or emigration, and the visits to the family gravesite on the mountain become less frequent. During this period of neglect, the kudzu, azalea, maiden grass, and wild smilax move into the circles of sunlight. The ornamental camellia and hibiscus that the family planted long ago are left unpruned, and the forsythia becomes feral. Saplings of red pine, lacquer trees, acacia, and Japanese snowbell take root and they are the frontline troops: the pioneer species of the mountain forest recolonizing its lost ground.

South Korea is a country about the size of Indiana and has a population of forty-eight million, one of the most densely populated nations in the world, and the dead keep piling up. According to the Korean Forest Service, approximately twenty million graves dot the countryside. I believe there are more than that; the dead are everywhere among the pines and boulders.

The graves of the Korean mountainscape can be somber places if you want them to be. Mostly, for me, they were a place of repose and rest, a place to sit down and eat an apple or, on a sunset hike as the pheasants roost in the pine boughs, to crack open a beer and watch city lights come alive at dusk. Down below, beer bottles were being opened and pork was being tossed on neighborhood grills. There was the sound of the children chasing the mosquito fogger truck and running into the toxic vapors. There was the loudspeaker sound of the vegetable truck peddler: *yangpa . . . hobak . . . mannul . . . gochu . . .* onions . . . pumpkins . . . garlic . . . peppers. Down below I could barely see my house. It was in an older neighborhood covered with magnolias, wisteria, ginkgoes, junipers, painted maples, paulowinia, fatsia, and zelkova. There were no yards, but many small vegetable gardens were surrounded by courtyard walls topped with shards of broken glass. The flat rooftops had bright yellow water tanks and clotheslines, and above those, the blood-red neon crosses of Christian churches flickered alive for the coming night. I borrowed these sacred places, for a quiet moment, from the family members who had put their loved ones into the earth. Blitz was always with me on these excursions.

We put up those birds, our daily venatic ramble. Hunting is possible in Korea, but the firearm laws and related regulations are so restrictive I decided it was not worth the effort. In his book *In Korean Wilds and Villages*, Sten Bergman wrote of a Japanese hunter living in Korea: "During his twenty-five years in Korea he reckoned that he had shot from 40,000 to 50,000 pheasants, shooting at least 1,500 every year." In Korean War literature, I've come across references to American soldiers gunning down ring-necks to add to their rations.

Blitz flushes so many pheasants those numbers seem very realistic. He consistently puts up pheasants for me to photograph every day. The partnership of the mission is strong: I read the crosshatch of mountains looking for gullies, the pockmarks of brushed-over graves, and islands of broadleaf trees among the sea of red pines. With our route mentally marked, I cast him out with hand signals into thickets of thorn-soaked briars and shadow-stands of bamboo. Out comes a cackling rooster and, rather than the boom of a shotgun, it is the *snick, snick, snick* of a camera.

One winter we came across a hundred acres of abandoned farmland and orchards near our home in Changwon. The government and construction company thugs kicked out the homesteaders and dug six-foot-deep trenches throughout the area that would be foundations for a high-rise ghetto of cookie-cutter concrete apartments. The livelihoods of the farmers were gone, and it would later be sealed over with asphalt. But a couple of

Fig. 30. Blitz near a hillside grave, one of millions throughout the countryside. The cocker spaniel flushed thousands of ring-necked pheasants during his short lifetime. Photo by James Card.

squatters held out, refusing to give up their farm plots, and the construction was halted. In the following year, the idle land grew wild and turned into the finest habitat any animal could ask for.

A typical day that winter went like this: As soon as Blitz sees that I'm putting on my brush pants and field jacket, he starts his whimpering and paws at the door. He waits for this moment like a coonhound waiting for sunset. The upholstery of the Hyundai beater is always covered with Spanish needles, beggar's ticks, mud, dog hair, and coffee stains. I park near a pile of trash and thrown-away appliances. The locals have discovered this land is a good place to dump things that cannot fit inside the government-issued garbage bags.

On the two-track leading into the fields, Blitz flushes a pair of Coturnix quail, bullet-like fliers that are smaller than the American bobwhite. We make our way to a ramshackle outhouse with a primitive squat toilet, one of the few buildings left standing. It is located in a tangle of wait-a-minute vines and satellite clumps of Eulalia grass. Blitz gets birdy and flushes three pheasant hens to my left, and to my right a water deer bolts off with hunch-backed bounds.

We make our way through trenches filled with rainwater and surrounded by tall grass. We jump migrating spotbills, mallards, widgeons, and teal. The ankle-deep creek attracts gray herons, and they fly away with a Mesozoic croak. Around the lowlands of the creek, groundwater bleeds into the soggy earth, and Blitz tears through the mud flushing a dozen snipe, which zigzag high into the air with bent wings. One day we flush a migrating Eurasian woodcock that stopped to probe for earthworms.

Blitz humps through the false indigo and Amur silver grass, bounding like a pouncing fox in high snow. At the edge of the forest, near a hedge of viburnum, he noses a pheasant kill. A small pile of breast feathers and one foot are all that are left. The inch-long heel spur is hardened with a dark shine. He was an old bird, a survivor, and we most likely flushed him a few times before. It was death from above. Goshawks are migrating through the peninsula.

I whistle my questing dog into a stand of cattails, and he disappears despite the orange vest he is wearing. There are rustling sounds, and out bursts a cackling get-up-and-go rooster pheasant. Already Blitz looks like a pin cushion from picking up hitchhiking seeds called Spanish needles or, as they are called in Korean, Goblin's needles (*Bidens bipinnata*, of the Aster family). I check his paws for the spiky burrs of giant ragweed, a plant Koreans call pig grass.

Near a collapsed farm building I named the Porn Shack (I found dis-carded manga comic books depicting cartoon characters sexually abusing each other), Blitz dives into a kudzu vine tangle to chase another rooster, screwed in tight against a stone wall of a terraced field. The dog circles back, and I reward him with a slice of *pyeonyuk*, a kind of headcheese. We hunt along these rock walls, which were built centuries ago. Along one wall is a hedgerow of overgrown *taeng-ja*. Known as trifoliate orange, it is Mother Nature's answer to concertina wire. I have donated blood to this tree many times. Its dagger-like thorns are pinpoint sharp, but it protects a fruit that is gastronomically useless. Its tiny orange is a sour fruit that might only be used for a bitter marmalade at best.

One of my proudest moments was when Blitz kicked the shit out of a raccoon dog. It was nearing sunset when we encountered the strange creature. In Japanese lore, the raccoon dog is a trickster spirit with huge testicles that it beats like drums, and it slings its giant scrotum over its shoulder like a backpack. It is half raccoon, half dog. It is the only member of its own genus, a weirdo branch of the canine family tree. It has the black bandit face mask of a North American raccoon, a bushy tail, rounded ears, and a gray and black brindled coat. Of its scientific name, *Nyctereutes* is the genus name, which means "night wanderer," and *procyoniodes* means "similar to a raccoon." It is nocturnal, omnivorous, and the only member of the canine family that will hibernate.

Another oddity is that it is the only member of the canine family that does not bark. When Blitz came upon it, it growled for a split second, but as it noticed the advancing spaniel, undeterred, was gaining speed, it crouched into a defensive posture. Blitz smashed into the raccoon dog and knocked it over like a defensive end leveling an unaware quarterback. They rolled over each other on the mountain slope. I grabbed a stick and drew my knife. It was not necessary. The spaniel kept up his attack and barked and snarled and lunged with pure aggression. The raccoon dog's body language immediately turned submissive, and the creature slunk away, tail tucked.

The winter days passed like this, and our sporting jaunts became a psychological crutch that helped me bear the bleak, drab, snowless winter of the Gyeongnam Province. Our outings saved me from the rubber room when the days grew short and the darkness came quickly. The spring led to the fly-fishing season, my guide schedule filled up, and sometimes Blitz came along, depending on the trip and clients. I was living in a floating world of outdoor pursuits, one of rambles with my dog after the world's most-loved game bird, and one of fly fishing for exotic fish few Western anglers have ever heard of. On a night when I was out of town, Blitz woke the house by hysterically barking at the door. Seizures took hold of him, and he went into shock. Our vet answered the call in the early morning hours and tried keeping him alive with an IV and injections. Blood tests later revealed there were no toxins. The vet theorizes that it was a "canine cardiovascular accident." Blitz had a stroke and was gone.

We went back to the mountains.

The last of the protesting farmers had been kicked out, and excavators had leveled the brushy fields that had been filled with pheasants, deer, ducks, and snipe months before. I carried Blitz up the mountain and, among the boulders and under a chestnut tree, I dug a deep hole with a trenching shovel. I wrapped him with his favorite flannel blanket, and I laid the tail feathers of roosters across his body. I covered the soil-tamped grave with stones and made a brush pile over it to discourage any scavengers. I climbed the chestnut tree and hung his collar from a high limb, and as I climbed down, I pruned away branches that could be used to climb back up.

I rested, smoked a cigarette, and sipped on a bottle of Munhak soju. The tears couldn't be wiped away fast enough. Farther down the mountain was the Buddhist temple where my wife asked the monks to recite their chants to guide the dead man's soul into another realm.

I'm not superstitious and tend not to believe in such phenomena or paranormal events, but this is what happened next: the pheasant roosters started calling out their signature "*kwak, kwak!*" One called from a ravine thirty feet away. Another one called from behind in the pines. Another called from across the valley. Another called from downhill, another uphill. Another called a few yards down trail. More calls from every distance and direction. I was suddenly surrounded by them, more than I've ever seen or heard. They came and called out, as if paying homage to the little gun dog that chased them but never killed them; they called out again and again and let all of the creatures of the mountain know that a hunter sleeps tonight. I sat there and mumbled many times, "We had a good run, we had a good run," and that later turned into, "I had a good run, I had a good run."

I loved the ruggedness of the Korean Peninsula and the way the mountains hide their secrets in lost valleys. I loved the freestone creeks that tumble down from jagged snowy slopes and meander to the sand beaches of the East Sea. I loved cold saltwater on my face and the sting of my spear piercing hard into fish flesh. I loved the weightlessness of a fly rod and the sharp stop of a backcast and the forward punch of an unfolding loop. I loved the gentle traitorous float of a dry fly as it drifts before a holding spot and the slash of silver that emerges from the deep. I loved the Eurasian widgeon, the Baikal teal, and the taiga bean geese that migrate to the peninsula on cold Siberian winds. I loved the tumbled scree slopes that pour off the mountainsides and the thickets of rhododendron and azalea that camouflage the deer and boars and pheasants.

I loved the *déjà visité* of knowing the Korean landscape yet experiencing it as new again and again. I loved the perennial bounty of orchards bearing citron oranges, chestnuts, and persimmons. I loved the smell of an acacia-wood campfire and the smell of thick-cut pork bellies and barbecued eel cooking on the grate. I loved the Korean dawn and studying mountains and rivers while most of the nation slept. I loved the eyes of all the raven-haired women, and I loved trading empty shot glasses with their brothers. I loved the woodcraft of ancient temple makers and the artists who paint scrolls of endless mountains and water. I loved wandering the narrow alleys of *hanok* houses and reading old poems of exiled hermits. I arrived in South Korea as a young man, and years later I was a husband and a father. I'd had a good run in a good land. It was time to move on.

15   Little Two-Named River

I became extremely depressed after I returned to the United States. I could not muster any urge to go fishing. I tried very hard to capture the zest for life that I had when chasing the mysterious trout in the rugged Korean mountains, but every attempt to cure this angling tristesse resulted in failure or even more depression. Everything in my life was going well—a beautiful wife and handsome son, a newly purchased house, a new job, a new pup, but my ambition to go fishing was shot.

I was working in Memphis as a magazine editor for Ducks Unlimited. It was a wonderful job with a world-class conservation organization, but the closest trout fishing in the mid-South region was in the Ozark Mountains of Arkansas about three hours away. I immediately started making overnight camping trips to the region. I attempted to recreate the magic I experienced in the mountains of Korea in the mountains of the Ozarks.

One day I parked at an access point on the Spring River, and I was eyeing up the water before me. There were two anglers: one guy decked out in waders, a ball cap, and a simple flannel shirt who was high-stick nymphing a stretch of current. And the other guy looked like he just stepped out of an Orvis fly shop after dropping a couple of grand. He had a talent for casting tailing loops.

An old sedan pulled in, and out poured a couple of kids and a woman wearing gray sweatpants and a sweatshirt. She lumbered over to the river with a spinning rod mounted with a Zebco 202 reel. She walked right into the river, soaking the sweatpants. Her kids threw rocks in the river while she pinned a baseball-sized bobber to the line and baited the hook with synthetic bait. Three more cars pulled up. Each vehicle had more rock-throwing children and mothers armed with bait and bobbers.. This was not the scene I was looking for.

In the 1940s, the U.S. Army Corps of Engineers dammed up the main rivers in this area, and ever since then, trout have been stocked in the cold tailwaters. More than 1.5 million trout are stocked every year in Arkansas. They are released into the rivers by the tens of thousands every month. A stocking schedule is published, and one can base angling endeavors on where and when the most trout are dumped.

The first time I was on the Spring River, I turned over rocks looking for clues on what kind of fly to use. I found the usual suspects of caddisfly, stonefly, and mayfly nymphs and some other aquatic insects. My fly box was loaded with all of the flies I used in Korea, and I was confident that most of these generic patterns would be suitable and presentable on any trout stream on Earth, including the Spring River. I experimented with a few different patterns and caught nothing. That was very strange.

Meanwhile, my partner, who had ridden up with me for a day of fishing, was catching them one after another. I finally yelled over to him and asked what he was using. "Egg," he replied. I shook my head and then nodded. And I asked if I could have one. I waded over to him and met him halfway. He gave me a peach-colored fly that was nothing but a round bit of fuzz with a hook in it. It was called the Y2K, which stood for Yarn, Two Knots. As in that was what was needed to make the thing: put the hook in the tying vice, bunch up the fuzz, tie two knots, snip the fuzz into a round shape, and done.

As a joke it was referred to as the State Fly of Arkansas. At the Spring River Fly Shop (store slogan: "Where there is no dam siren"), they sold the same kind of fly but more of a brownish color to resemble the fish food pellets the trout are reared on at the hatchery.

I was laid off from my new job after only a couple of years. At first, I felt bitter and betrayed, but I later realized that my time in Memphis was a temporary stepping stone that provided a useful toehold for me to get reestablished on American soil and give me the footing I needed to get to a better place.

The better place ended up being central Wisconsin. A World War II veteran, who had served under General George Patton, had come back after the war and worked as a carpenter. In 1952 he launched a coin collecting newsletter on his kitchen table in Iola, Wisconsin. By the millennium, his company was the world's largest publisher for hobbyists. They published books and magazines on all kinds of subjects, ranging from coin collecting to antique cars. They had an outdoor division that published literature on guns, hunting, fishing, trapping, and survival. That's where I fit in.

I did not make the same mistake twice: when relocating your entire family across the world or across the country, always rent first, preferably a cozy but affordable dive. I found a ramshackle ranch house tucked in the woods outside of town.

Meanwhile, I sought to keep a connection to fly fishing. I still went out but it felt like a chore, and the passion for it that I had in Korea was gone. I assuaged this for a while with smallmouth bass and later brook trout, the prettiest native fish around, and that worked for a while.

Also, as a father of a young son, my focus shifted to teaching him to fish while putting food on the table. Catch and release is appropriate in certain waters, but if you can bring home a mess of perch, bluegill, and crappie and put on a fish fry and save a few bucks in the process, all the better. I started him out simple, and we used bobbers, sinkers, and hooked nightcrawlers and fished out of a canoe. I loved this very much. All of the deep study of fly fishing was gone. All the technical mastery and obscure knowledge disappeared into one phrase: pass me the worms.

Once, while he was bobber fishing on shore, I cast around with my fly rod and found a bunch of bluegills near their spawning beds. I observed that there was about a three second delay from when the foam beetle hit the water to when they smashed it. It was enough time for me to pass the rod to my son so he could feel what it was like to land a fish on the long rod.

I put the fly down and quickly passed the rod to him. At this age— around six—the only thing he knew how to do when a fish struck was to set the hook and furiously crank a spinning reel. There was slack line on the ground, and when he set the hook and reeled, he felt nothing but slack. Not understanding what was happening, he gripped the line against the rod and ran backward for thirty feet until the bluegill was dragged to shore.

We spent many hours in the canoe and in kayaks. All of the fishing tackle I had accumulated over my lifetime had been stored in my parents' basement when I was in Korea. I gave it all to him: every crankbait, spinner, hook, bobber, and jig. I taught him how to fly fish but did not push it on him. One year he got a beginner fly-fishing kit for Christmas. One thing I learned as a guide is that high-end fly rods are wasted on beginners. Let them cast a broomstick for a while until they develop the insight to look for a finer rod. My fly-fishing gear I kept in a special drawer in a filing cabinet. I kept two fly rods strung up and ready for immediate use in my garage. They hung on deer-antler pegs, and I rarely touched them.

Another factor that kept me away from the fly rod was the bow, rifle, and shotgun. During my time in Korea, I did not hunt and now was time to make up for it. I went after whitetail deer, wild turkeys, ducks, geese, ruffed grouse, woodcock, and ring-necked pheasants. I updated my gear and guns and spent hours on the range dialing in the perfect zero. My instinctive archery skills had atrophied so much that I could barely hit the bullseye at ten yards. I semi-retired my take-down recurve and got a compound bow with a peep sight, and I shot with a trigger release.

My wingshooting skills were so rusty I could not hit a blimp moving slowly across the sky. I self-analyzed my form and performed a daily routine of mounting the shotgun to my shoulder and tracking the ceiling corner seam in my office. I spent hours shooting skeet, and I would not advance from one station to the next if I had not mastered that shot or unless I had an idea of what was tripping me up. The new pup was an American water spaniel, and I spent hours working on obedience and retrieves to train him into a gun dog.

I also embraced ice fishing, a long-cherished sport that I had abandoned living in the warmer climes of Korea's southern coast and Memphis (both are along the thirty-fifth parallel). Ice fishing is the antithesis of fly fishing. The rods are short, and there is no casting whatsoever. It is cold, tough, and sometimes brutal. If fly fishing is a gentleman's pursuit, ice fishing is the enterprise of boreal barbarians. Its roots are more about survival than sport. I've always enjoyed fly fishing along snowy banks on a winter day and working some midges over small riffles, but that went out the window once I was back in the land of frozen lakes. I took to the hardwater with a vigor that can be created only by breathing in frosty air.

The urge to recreate or recapture my fly-fishing experience in South Korea still lingered. Nothing seemed to excite me despite living in Wisconsin: a land of over 10,000 lakes, with 3,000 trout streams and bordered by two Great Lakes and the mighty Mississippi. I missed the rugged misty mountains of Korea and the freestone streams that tumbled through their hidden valleys.

What I needed were mountains, and the closest thing to mountains in the upper Midwest is the Driftless Area. Not exactly mountains but steep, giant bluffs that form hundreds of coulees where spring creeks emerge. When we visited my family in the area, my wife always commented that it reminded her of Korea. The area is formed of karst geology, much like the Samcheok area where Sohan Creek flowed from the maw of that mysterious dark-energy cave. If I could find a creek like that, I could shake my blues. What I needed was to fish a creek that flowed out of a cave.

On a winter day, I drove to Coldwater Creek in northeastern Iowa. In the early autumn of 1967, Steve Barnett stood before a trout stream that disappeared into the base of a 150-foot limestone cliff. He swam under the rock lip and surfaced in a room that was over sixteen feet high.

On subsequent trips, he and his college buddies navigated a series of underwater passages with primitive scuba gear. The cave opened up into huge hallways with stalactites and dome-shaped rooms dripping with flowstone. Coldwater Cave is still being explored, and seventeen miles have been surveyed—and flowing through it is one of the state's best trout streams, Coldwater Creek.

Southeastern Minnesota and northeastern Iowa are part of the Driftless Area, the spring creek country that was left untouched by the last glaciation. The region is a limestone landscape that tends to hold many caves, sinkholes, and springs. In this border country there are caverns so large you can drive a truck through them. In some you can walk for miles without stooping. There are underground rooms so big a house could be built inside. There are waterfalls only a handful of humans have ever seen, and the waterfalls are fed by rivulets that are so small only a trout could pass through. There are muddy belly-crawl passages still unexplored that may drop into gaping canyons. And then there is the Odessa System, a theoretical cave network running through the area that is thought to be bigger than any of the caves in the region. It may contain a subterranean river so powerful that it might be impossible to explore without someone getting killed.

One of the reasons the caves are slow to be explored is the same reason anglers think twice about fishing: high water. In spring and summer, runoff and steady rainfall turn the underground rivers into flooded pipes. There is a story of one caver sitting down to take a break. After a few bites of his sandwich, he looked down to see his lunch floating away. The team made it back to safety with their faces kissing the rock ceiling for the last half inch of airspace.

I hiked through knee-deep snow and found it easier to wade in the creek with my knee-high rubber boots. The closer I got to the cave the more caddis casings freckled the river stones. Brook, brown, and rainbow trout are present in the watershed, and I spotted some as they darted away from my upstream sloshing.

The creek runs against the wall of a limestone cliff after tumbling over a gravel cataract, and as you approach the cave—there it is, a glassy pool at the base of a rock alcove. It seems to ooze underneath the bluff. I spotted some movement in the still water. A nine-inch rainbow finned about, lazy and unconcerned. I crept closer and crouched down and looked as far as I could into the back of the cave. It was all water and darkness and more trout. I counted six. There were probably more back there.

In 1969 local Iowa cavers installed a locked gate across the underwater entrance and passed the keys to the Iowa Conservation Commission. Now the cave can be accessed through two drilled shafts: one drilled by the state of Iowa but now under the control of a local landowner, and another by John Ackerman, a private cave conservationist.

I stepped into the pool and shined a flashlight into the cave. The trout disappeared into the depths, frightened by the powerful beam. I waited a long time for them to return, but for all I knew they could have been miles away. This was a beautiful stream. Any angler would deeply appreciate it. It checked every box. I made only a handful of casts and caught one rainbow. I could have caught many more, but I knew that wasn't the point. I was glad I had made the trip, but it did not satisfy me or recreate the experience of fly fishing the karst kingdom of the Samcheok area. I drove home pleased with the day's outing, but the angling ennui was still there.

After a few years of getting established, we got the itch to buy a house of our own. I took a gamble and bought a foreclosed ranch house at a real estate auction. The bids went up in $10,000 increments, and within sixty seconds I was a homeowner. Not just any kind of homeowner, but one with waterfront property. The Tomorrow-Waupaca River flowed through the backyard, and a lifelong dream came true: to own land along a trout stream.

I immediately had visions of fly fishing every day and getting back into the rhythm of fly casting and awakening that old muscle memory. Not so fast. The previous owner had let the place go to hell. The river could barely be seen through the invasive brush: an ugly and undying mix of buckthorn, Russian olive, and honeysuckle.

Inside the house, dogs and cats had been free to piss as they pleased. Weird wallpaper adorned the walls. There were trailer loads of junk that needed to be hauled off. The basement was a moldy dungeon. Amid the invasive brush choking the woods, junk and litter was scattered everywhere. The lawn was a knee-high meadow of weeds. Any idea of going fly fishing in my new backyard—even if only for a few minutes—seemed like a frivolous waste of time.

With my Jim Gem dibble, I planted hundreds of Norway spruce, red pine, white pine, crab apple, silver maple, hazelnut, and tamarack every spring. I repaired the boat dock, and when that was smashed apart and sent downriver, I built another one. I cut, split, and stacked endless cords of firewood to feed the woodstove. I converted a chicken coop shed into a writer's cabin. In a strip of woods I nicknamed Shooter's Alley, I built an archery and firearm range. Tree stands and blinds for deer and turkey hunting were erected in the woods.

I nicknamed the place the fifteenth Boulder Rod & Gun Club, after Ryōan-ji, the Zen Buddhist temple in north Kyoto, Japan. The temple has an expanse of raked gravel, and situated in the yard are fifteen boulders of different shapes and sizes. From any point in the viewing area, a person can see only fourteen of the fifteen stones. Nobody knows if this was intentional or an accident. It is said by the monks that if one were to see all fifteen boulders, one would have achieved enlightenment.

I interpreted this based on my time spent in the field. It is the same kind of enlightenment as when I intuitively knew where trout would be positioned in a stream and was never wrong. With the same kind of meditative reflection the monks dedicated to the Ryōan-ji rocks, I studied the boulders of the Jirisan streams and the trout that existed among them. Finding the fifteenth boulder is when a hunter studies a hillside for hours, and the small brown rock that he has glossed over many times before is revealed to be a bedded deer. The fifteenth boulder is when an angler approaches a set of riffles and knows within an inch of where to cast and present the fly—but only after thousands of hours of observation. The enlightenment is then confirmed by a trout rising to take that fly: that is the fifteenth boulder. It is the a-ha moment, it is when the light bulb turns on. Since ancient times, it is the mental state sought by artists, inventors, entrepreneurs, clergy, writers, designers, craftsmen, and farmers. It is not when man discovered fire. It is when he learned how to make fire.

When reflecting on my self-invented rod & gun club, I think there was more gun than rod. My urge to fish the river in any serious manner did not exist. The river was always there and always on my mind. I stared at it every morning while having a cup of coffee and while sipping on an evening beer. I kept a journal of the comings and goings of wood ducks and hooded mergansers. I kayaked different lengths of it and fly-fished a few spots here and there.

But I hardly touched my fly rod. Instead, I reached for the chainsaw, the rifle, the shotgun, the bow, the shovel, the rake, the hammer, the drill, the crowbar, the paintbrush, the bucket, the scythe, or the axe. Or the fungo bat, as by this time I was my son's Little League coach.

In this section of the river there were only brown trout and a few species of minnows and suckers. I found this to be exceptionally dull. In my garage workshop and in my cabin and office, I hung photos of the exotic trout I had caught in Asia. I missed them. A river with only brown trout seemed so mundane.

The river was fished and explored by one of America's early outdoor writers, a fly-fishing preacher by the name of Onnie Warren Smith, who wrote under the byline of O. W. Smith. He was born in 1872 in Weyauwega, Wisconsin—a small town that I would later cover as a journalist—and he fished the Waupaca River when he was a kid. He studied at Lawrence College in Appleton and went on to serve congregations around the state in Oconomowoc, Mondovi, Washburn, and other places. He started out as a Methodist and later became a Congregationalist.

Smith was a lover of angling and literature and traveled with a truckload of books. He wrote for *Outdoor Life* and other sporting magazines throughout his life. He authored *Trout Lore*, a bestselling how-to book on trout fishing, and his best work was a literary work published a year after his death in 1941 called *Musings of an Angler*. He also wrote books on castings and tackle, northern pike, campfire tales, fishing with live bait, and fly tying.

In the August 10, 1912, issue of *Forest & Stream* magazine (the predecessor to *Field & Stream*), he wrote an article about the river titled "To-morrow and the Day After." It starts off by sorting through the confusion of trying to locate a river that went by two names. "A rumor of good fishing to be found in a river called 'To-morrow' had reached me and be it said that name was new to me. When I asked my wife if she had ever heard of a stream with such an uncertain sounding name, she quoted, 'To-morow never comes,'" he wrote.

Smith tried to find the stream but could not locate it on a map, yet he kept hearing references to other anglers catching trout there. Finally, a pen pal mentioned in a letter that he "took a fine catch of trout" from Tomorrow. Smith immediately replied and later got an answer: "To-morrow is the English translation of the Indian word Waupaca, and around here, it is always applied to the Waupaca River."

He then sorted out how the river had come to have two different names. He learned from a local legend that Native Americans canoeing the area would portage from the Wisconsin River in Stevens Point to the Waupaca River and, from there, canoe down the Waupaca to the Wolf River, cruise through the Lake Winnebago system, and pick up the Fox River to Green Bay. The portage would involve a day of hiking and hauling, and when settlers asked them how far they were traveling, they would reply, "Waupaca," meaning by tomorrow they would have their canoes back in the water.

My interest in the river's brown trout was stirred only when my son started fishing on his own. I taught him the basics of fly casting, but by this time he was catching lots of brown trout with Panther Martins with a light spinning rod. I was very happy for him. The fly fishing could come later. We found the remains of an old wooden boat with pieces of ribs attached to the planking with brass screws, and I surmised it was old enough to be from the days when O. W. Smith was fishing the river. Meanwhile, I kept working on the property. There was always more work to be done.

My son disappeared for hours at a time and came home with reports of the brown trout that he caught. Lots of small ones, under twelve inches. He practiced catch and release although a few times the hook was embedded deeply in the fish and blood was running from its gills, and I taught him how to clean and cook his catch. The trout he caught kept getting bigger—now into the teens. Some thirteen-inchers, fifteen-inchers, seventeen-inchers. Trout of that length would have been considered trophies on many of the Korean creeks that I fished. It was when he started catching the occasional brown trout over twenty inches that my interest was piqued. They were magnificent specimens—hard fighters and cunning survivors.

It was around this time that I let the angling ennui die. It was time to let the black dog of depression wander off into the woods and go fly fishing again. If the river that I now lived along had only brown trout to offer me, I would devote my interest to the species of trout that gave birth to the sport of fly fishing. On the first page of a small notebook, I wrote: "The Brown Trout Study." There was still work to do, but I would give myself permission to grab my fly rod and vest and go fishing on a small river with two names.

16   Ninth Station Coda

Last summer I spent many nights at a ballpark in Plover, Wisconsin. My son always had to arrive early to warm up before the baseball game, which left me with some time to kill. A few minutes away was the Korean War Veterans Memorial. A path leads through a park of mature white pines. On each side of the path are alternating posts. Each post is topped with a small metal sign that explains the symbolism of each triangular fold when folding an American flag. Owen Helgeson of Troop 298 put this together as an Eagle Scout project.

Station 1 explains that there are thirteen folds in an American flag and each fold represents one of the original thirteen American colonies. After that, each station and fold of the flag is a remembrance of past wars.

Station 9 reads: "For the brave men and women of our armed forces who served and died on the beaches, fields and mountains of Korea (1950–1953), we salute you."

The trail merges with a causeway sidewalk extending to a small pine tree–covered island in the middle of Lake Pacawa, a man-made lake with riprap along the shoreline. It very much looks like a reservoir where I once bass fished in South Korea. I found this to be ironic: South Korea, a country in the process of building more dams to add to the thousands of artificial reservoirs that already exist, and Wisconsin, a state of thousands of natural lakes and a leader in dam removal and river restoration.

On the island there is a huge stone wall and on it is engraved:

KOREAN WAR

JUNE 25, 1950–JULY 27, 1953

132,000 WISCONSIN'S SONS AND DAUGHTERS

SERVED THE CAUSE OF FREEDOM.

801 KILLED IN ACTION.

4,286 WOUNDED, SICK OR INJURED IN ACTION

111 PRISONERS OF WAR, 54 DIED IN POW CAMPS

84 STILL MISSING IN ACTION

There is bronze statuary of five figures looking off into the distance. A nearby plaque explains it: "These statues face Korea in the Far East, in memory of all who lost their lives, and for those still missing in action. They came from different walks of life to help defend freedom in a country that was unknown to many. A Nurse leads the group, because without the dedication, compassion, and bravery of the medics, many would have not returned. Next are Army and Marine Infantrymen, because the infantry bears the brunt of the battle. Then comes an Air Force Pilot and a Sailor, without whose support, the battle would soon have been lost. There are no weapons in this memorial. This is to confirm the veteran's wish for peace. All five figures are touching, symbolizing the brotherhood that comes from battle. They stand in memory of all veterans who served our state."

The statues certainly face east, toward New York and Europe, but not the Far East. Korea is in the opposite direction, toward California—the same trans-Pacific route that all soldiers took to get to the battlefront in 1950. If anything, it would appear the statues had their backs turned to distant Korean shores—the opposite of the symbolism the memorial is trying to convey. For me, a lover of geography and compass points, it was an ignorant yet expensive mistake. Most visitors probably never notice.

I spent most of my time at the other walls reading over 770 memorial tiles that family members or friends put up to honor loved ones who served in Korea.

There was Melvin Handrich of Manawa, a Medal of Honor recipient. It was awarded to him for his bravery on Seobuk Mountain—a mountain a few miles away from where I once lived in Changwon. On August 25 and 26,

1950, he held off overwhelming enemy forces until his forward position was taken and he was killed. When the position was retaken, over seventy dead North Korean soldiers were counted—all presumably killed by Handrich.

There was Harold E. Sitler, who was wounded on November 29, 1950, and left for dead at the Battle of the Chosin Reservoir. He was held as a prisoner of war by Chinese soldiers for 1,009 days.

There was Corporal John Reese Jones of Wild Rose, who was stationed at the Eighth Army Headquarters in Panmunjom and cooked meals for Walter Cronkite.

There was Second Lieutenant Jerome Sudut, a Medal of Honor recipient from Wausau. It was awarded to him for his actions at Heartbreak Ridge. With a submachine gun, pistol, and grenades, he charged through enemy fire to a bunker and killed three hostiles and scattered the rest. He reorganized his platoon but, meanwhile, the enemy slipped back into the bunker via trenches and fired upon them. He charged in with a rifleman, and when the rifleman was wounded, Sudut grabbed his rifle and killed three of the new occupants. One enemy was left inside, and he jumped into the bunker and killed him with a combat knife.

There was Alvin L. Joyner, a USMC corpsman who chose not to carry a weapon. He was the first conscientious objector to be awarded a Silver Star, along with a Bronze Star and three Purple Hearts.

There was Master Sergeant Howard K. Inderdahl of the first Calvary Division. He was awarded a Bronze Star with Valor. He survived the war and returned to Scandinavia, Wisconsin, and authored *In the Hills of Korea*, a book that can be found at local libraries.

There was Private First Class Harland Thoen. In 1947 he was declared the U.S. Army's lightweight boxing champion in a fight refereed by Joe Louis. On July 24, 1950, he was listed as missing in action in Korea.

There was Medal of Honor recipient Corporal Einar Ingman Jr., who attacked a machine gun emplacement during the Third Battle of Wonju. He lobbed in a grenade and finished off the rest with his rifle. As he approached a second machine gun position, part of his left ear was blown off and he was shot in the face. He got back up and killed the gun crew with his rifle and bayonet. He settled in Tomahawk, got married, and had seven kids.

And there are so many more. This book is dedicated to those who fought to create a nation of free people on the Korean Peninsula.

Postscript

THE CODE OF THE EXILED ANGLER

For many years I kept a pocket-sized book in my fly vest. It was *The Fisherman's Calendar* by Yun Sondo. Yun is one of Korea's greatest poets, but he was also a rebel, a teacher, and a man of nature. He was born in Seoul in 1587 and rose through the bureaucratic ranks during the Choson Dynasty. He was banished for speaking the truth and criticizing the power hierarchy. He died in 1671, and most of his life was spent in the country, studying nature and writing and fishing.

His masterpiece is *The Fisherman's Calendar*, and the *sijo* poem is divided by the seasons with each season having ten verses. In the third line of every verse is a refrain: *Chigukchong, chigukchong, oshwa!* It is an onomatopoetic line that is to sound like the clunking and pulling of an anchor chain and the swept sound of an oar through the water.

The poetry was translated by Kevin O'Rourke, a professor at Kyung Hee University. He came to Korea from Ireland in 1964 as a Catholic missionary, and he later earned a PhD in Korean literature from Yonsei University. He is one of the most prolific Korean-to-English translators in history. My copy of *The Fisherman's Calendar* was published in 2001 by Eastward Publications in Seoul.

In the classical art and poetry of northern Asia, the fisherman is an archetype of a rusticating wise man who lives simply and close to the earth and water. The lifestyle is purposeful, and Yun Sondo chose to describe the life of a fisherman to create his magnum opus.

Angling is an act of rebellious escape. It always has been. On the commercial side, fishing has always attracted those who choose to trade the comfortable salary of a straight job for the risk, freedom, and rewards that the water might offer up. Better to toil on your own boat and be the captain of your destiny. This is an ancient archetype. In recent times this was portrayed by unlucky Santiago as he prepared his tackle and talked about the great Joe DiMaggio before heading far out into the Gulf Stream. The archetype is embodied by Quint when he fired his Greener Light Harpoon rifle into the man-eating great white shark that terrorized Amity Island.

On the recreational side, fishing literature is rife with references to "going fishing and getting away from it all." Yun hints at this urge throughout the poem. He writes about being far away "from the dusty world of men" and the pleasure of getting far away from the world of civilized society. Yun notes the royal court is too entrenched in their cities to appreciate the natural beauty that he can see, and he asks himself if he is somewhere between Earth and Heaven. By living and fishing in a remote area, he contends he is "doubling my joy with distance."

With angling it is important to develop a high level of skill and knowledge but also to not take yourself or angling too seriously. In older times, catching some fish meant your family would not starve. Those stakes were very high. The pressure was on: it was either hook set or hunger. At the same time, the anglers of old that Yun describes caught fish but enjoyed the simple pleasures that nature offered. They saw their existence on the water as part of a bigger picture. They marveled at the colors of sunsets, willow groves, the paths among pines and rocks, autumn colors and orioles, seagulls, cuckoos, wild geese, and crows. They appreciated the warmth of a winter sun. They took time to appreciate the small things, and this often gets lost with modern anglers, who get caught up in materialism and status.

Some fly anglers are obsessed with wearing their vintage Patagonia wading jackets from the 1980s, donning a herringbone sport coat at the lodge, and having the newest rods and reels. Narcissus fell in love with his own reflection in a pool of water, and many fly anglers do the same on trout streams.

There are some fly angling snobs who believe upstream dry-fly presentations are of the highest order. Here is a secret: dry-fly fishing is one of the easiest forms of angling out there. A person dumber than a bag of hammers can figure it out very quickly. Subsurface presentations are more difficult as you must visualize, in your mind's eye, how the fly is moving underwater. It is like the difference between watching a baseball game on TV and listening to Bob Uecker on the radio and having to form a streaming image of the play action in your head.

There are fly anglers who look for outside validation in some form or another. Once upon a time I decided I wanted to be an undisputed master fly caster. I figured that becoming a certified fly-casting instructor would be a good benchmark. One must pass a skills test and an oral exam. I found the Fly Fishers International testing criteria, and for three years I practiced casting in a grassy park in Gwangyang and replicated their skills test. Once I had expertly passed every metric on my own, I no longer cared about this absurd credential. When you must pay an outside entity a large sum of money for a stamp of approval to show that you are good enough when you already know you are good enough, you are the sucker.

There are small details that indicate Yun spent some time with a fishing rod in his hands. Yun writes of rods being rigged and readied to cast. He notes big fish that swim near the surface and other fish sighted in deep water. He observes murky water running clear after the end of a long rain. He talks of leaving one fishing spot to find another. He quotes a Chinese poet who was writing about fishing three hundred years before the birth of Christ. The fisherman in his poem wears a bamboo hat, a hermit cape, and a raincoat made of green rushes, and he eats a simple snack of rice wrapped in lotus leaves. He writes of his wicker creel and building a reed fire and broiling the fish, one by one, and filling his gourd cup.

He writes of anglers contending for the best spots and then giving the best spot to a white-haired old man out of respect. Yun writes about making the best of situations, and he mentions he might catch a man-souled fish—based on an idea that the souls of drowned men are passed on to fish. Throughout the long poem he makes subtle references to Taoism, ancient legends, and military battles. Like all anglers before and now and forever, he is annoyed by mosquitoes and blowflies.

But he sure as hell does not take himself too seriously. He sings boat songs and writes of passing out drunk and drifting into rough water. He wakes up to see peach petals floating on the surface and thinks paradise is near. He mentions misplaced wine bottles and taking naps under the boat tarp. He notes the morning mist on the river and the warming of a new day. There are gentle waves that "glide like oil," and trees that talk in the wind. Yun writes of the ultimate achievement for any angler in any century or any county: a day so joyful he forgets the setting of the sun.

TEN RULES FOR CREATING A SELF-MADE EXPATRIATE LIFE

Carving out a life in a foreign country where you do not speak the language is one of the hardest undertakings a person will ever experience. It is more difficult if you have minimal funds and sketchy employment prospects. These rules are not for the diplomatic corps or corporate executives or military personnel who live in protective bubbles. These rules are for those who live outside the walls and blaze their own path. These rules are for those who operate without a safety net in strange, distant lands. These rules are for those who can say: "All I've got is a pair of balls." *Gajin geonbulal du jjok.*

1. DEVELOP LITERACY AND FLUENCY

First learn the phrases of survival (bathroom, beer) and courtesy (please, thank you). Some languages are more difficult than others, so accept that and move ahead. Korean is a simple phonetic alphabet, and a basic understanding can be figured out in a day. After that, reading is about practice and memorization. Use flashcards to develop vocabulary and carry them everywhere to practice in spare moments. Your ability to read is judged by how quickly you can read road signs in a fast-moving vehicle. It should be an automatic reflex. No more sounding out words. It must be instantaneous.

The world is your language lesson. Practice basic conversation wherever you go. Buy textbooks and work through the grammar. Get a tutor, preferably an attractive one who can also be your lover. The goal is to develop reading, writing, and fluency skills so that you can function on your own without the help of a native speaker acting on your behalf.

Without being able to read, write, and speak Korean, I would have never been able to do any fishing of consequence. I would have bummed from one random river to the next. I would not have been able to do the intensive mapwork to locate the hidden valleys and backroads. I would have never been able to find clues such as a sign advertising fresh trout sashimi (which means the trout are farmed and, if they are farmed, they can escape, and if they can escape, they can form wild populations in nearby creeks). I would have never been able to scour hundreds of pages of scientific literature written in Korean concerning the peninsula's freshwater fish. I never would have been able to connect any dots. I never would have been able to get tips from local mountain people who generously offered information that flowed with cups of rice wine.

There comes a time when the language studies become too advanced, and it is a matter of opportunity cost: the intellectual energy of language study could be better invested elsewhere. This is when the novice decides to turn pro. Usually deciding to take your foreign language ability to the next level is influenced by careerism: resume building, university admissions, professional development, academic work in area studies, translation work, and promotions, for example. Once I decided that none of those things were of any concern to me, I focused on studying Korean to live a better all-around life and to use that skill to learn more about things that I was passionate about.

2. BECOME MOBILE

I had an international driver's license issued by the American Automobile Association (AAA) before I went to South Korea. I later learned that international driver's licenses are a scam. Anyone can issue them, including my friend Johnny Ham, a Canadian *kyopo* who printed them on his own in his basement home office in Daechi-dong—the same kind of below-ground apartment that gets flooded in the film *Parasite*.

Get a state-issued driver's license as soon as possible. Study and do whatever is required to earn this. Get motorized transport. It could be a scooter or a motorcycle, but a car is better and a truck is the best. The public transportation system in South Korea is spectacular but, even so, this extremely limits the country one can explore. Most expatriates in South Korea are slaves to public transportation and taxis. They never get off the beaten path (although they think they are leading adventurous lives). Some never get wheels because of frugality or practical reasons or because of a hipster minimalist attitude toward owning expensive possessions like a vehicle while living abroad. These people tend to be moochers and should always be shunned. How many minimalists does it take to change a light bulb? One, but the son of a bitch wants to borrow your stepladder.

There are no ways to effectively access the backcountry of a nation without your own motorized transport (unless by horseback or water-craft or bush plane). In the early days, I tried to use public transportation to visit some remote areas. An express bus will get you to the main city of the region. The bus that makes a circuit into outlying small towns will get you closer. The bus that runs into the rural country that you want to explore usually passes through only a couple of times a day. Much time is spent navigating bus schedules and waiting for the desired bus to show up and take off. By the time you are dropped off in the backwoods, when the walking begins, your weekend is half over and there is a growing paranoia about how you are going to renavigate the bus schedule in reverse to get back home. With a vehicle you can depart in the middle of the night to make it to a remote destination by sunrise. You can maximize your time spent in the backcountry and spend less time in transit. Packed with gear and provisions, the vehicle becomes a mobile base camp.

3. STUDY THE HISTORY

The new country you are in will seem baffling at first, but there are reasons for everything. To make sense of it you must understand the history of the land. There are reasons why some cultures grow rice and others grow wheat. There are reasons why this city evolved next to a river and another at a deep-water port. Starting from prehistoric times and reading until the modern era can be daunting. At minimum, learn as much as a high school student would be expected to know and then study the key turning points in history that forged the nation's identity. From there read about the history of whatever interests you personally. A serious student of martial arts will read about warfare, and a gourmand will learn about agriculture. To be a better outdoorsman, I studied how Koreans interacted with the land during the old days—forestry, farming, hunting and gathering, fishing, geology, carpentry, and dam building—even visiting Byeokgolje in Gimje. It is the oldest artificial reservoir in the country, built in AD 330 by King Biryou of the Baekje Kingdom. Visiting this ancient reservoir helped me gain a historical perspective on the pathological dam building during modern times. The best modern travel writer who blends history into his writing in order to understand what is currently happening around him is Robert D. Kaplan.

4. FIND VALUE IN SOMETHING THAT
THE NATIVES HAVE OVERLOOKED

People take things for granted. It is a universal human characteristic. As an expatriate you are an outsider looking at a new world with a fresh, curious (and often naive) perspective. I think of Alex Kerr's book *Lost Japan*. He describes finding antiques in his adopted homeland in the storehouses of family homes and purchasing centuries-old calligraphy for very small sums. He noted that in Japan's rush to modernize and surround themselves with shiny new consumer goods, their artifacts and antiques were treated as junk from a bygone era. He later became a prominent art and antique dealer.

In South Korea, my friend Robert Neff found a passion for studying Westerners in Korea during the late Joseon Dynasty. He is probably the most knowledgeable person in the world in that obscure subject area. There are plenty of Korean scholars who are experts in the Joseon Dynasty, but few have looked into the activities of the adventurous Westerners who arrived in the Hermit Kingdom. The last I heard, he had amassed the largest private collection of black-and-white photographs from that period.

For Steve Karsen, a science teacher at a Christian school, he hunted down and handled every reptile and amphibian on the Korean Peninsula. He then discovered a rare salamander the world had never seen before, and it was named after him. Through his passion for herpetology, he poked around in remote places that no Korean wildlife biologist had ever thought to look.

For David Mason, it was his curiosity about Korea's native shamanistic religion of the mountain spirit that led him to study it extensively and publish the first book in English about it. Buddhist temples get all of the attention, and this was an obscure religion that nobody—Korean or anyone else—had ever thought to share with the rest of the world.

For Thomas Duvernay, it was mastering *goongdo* (the way of the bow), an obscure martial art even in Korea. When he hit a target that was 159 yards away with five consecutive shots, he got a perfect score and earned his *muho*, or archer's pen name. During a ritual ceremony, the director of Tiger Forest Pavilion gave him the name Cheong-ho, meaning Blue Lake, a reference to his Michigan background. Duvernay produced and directed a documentary about horn bow construction and wrote a book about traditional Korean archery.

For me it was my quest to discover overlooked populations of freshwater fish that offered an exceptional sporting experience. These wonderful and unique native fish were there all along and have been caught by Koreans for centuries with nets, traps, baited hooks, and every other contraption—but not with fly rods.

5. FITNESS IS ESSENTIAL

Expatriates can let themselves go. Alcoholism, overeating, sleep deprivation, and Falstaffian debauchery are hard on the body. In some countries the local cuisine is delicious and inexpensive; it is easy to pack on the weight. It is often easier and cheaper to eat out than to prepare meals yourself. In South Korea, there is no bar time and one can drink until sunrise and amble off to bed like a bent vampire to sleep away the day—a day that could have been spent fishing. South Korea was an early adopter of high-speed internet, and that makes it way too easy to surf the web for the next dopamine fix.

One discipline that kept the vices in check for me was regular exercise and constant strenuous activity outdoors. During my first year, I bought a pair of fifteen-kilogram (thirty-three-pound) dumbbells and developed a weekly indoor regimen. I supplemented that with outdoor workouts at playgrounds: pull-ups, dips, hanging leg raises, and such. The always-nearby mountains became my gym, and I had an "empty up, full down" routine: the backpack was empty hiking up the trail, and on the way down I loaded it with chunks of wood that would later be used on my barbecue grill. I often hiked with a Burton Bar, a heavy steel bar that I found in an abandoned coal mine in the Taebaek Mountains. After reading a biography of Sir Richard Burton, I learned that to stay in shape he took walks with a large iron staff.

6. BE A CAREFUL ACCOUNTANT AND A CURIOUS ANTHROPOLOGIST

Being broke in a foreign country is a miserable existence. Read Orwell's *Down and Out in Paris and London* for notes on that. I met many deadbeats in South Korea. I met a few hustlers reneging on their student loans and others dodging child support payments and alimony. In South Korea, paychecks are paid out once per month, so one must budget with forethought.

While handling the currency of your new country, there is a sensation that it is like Monopoly money. This allows you to make financial decisions in a neutral or insane mindset. As an expatriate, you most likely have low expenses and obligations, and that frees up more brain power to focus on other matters, such as learning how the locals handle and invest their money. I was regularly astounded by how many Koreans fall for endless get-rich-quick schemes and swindles. There is a culture of bribery on all levels of society, and it gets interesting when you get personally involved. (Giving a teacher a small gift is common, but rich parents of dumb kids are extra generous.) You develop a praxeology of following the money and seeing who benefits. On a national scale, I saw ghost airports built that didn't serve any purpose (Uljin, Daegu, Yangyang, Muan), and on a small entrepreneurial scale, I always scratched my head at the clusters of copycat businesses that set up shop right next to each other.

7. HAVE A PORTABLE PSYCHIC ANCHOR

An off-duty policeman might not leave his home without a sidearm. A pastor might not leave the parsonage without a Bible. A poet might not go outside without a pocket notebook. A gambler might not head to the casino without his lucky charm. A married man working on the road starts the day with a photo of his family in his wallet. For me in South Korea it was a Leatherman PST II.

It was with me when I first arrived in South Korea, and it was with me when I left. It never left my side. Once while in Incheon International Airport, I forgot I had it on my belt and I made it through the full-body x-ray scanners. After that, I had a layover in Narita and I passed through those security checkpoints, and somewhere over the north Pacific I noticed the leather sheath digging into my side.

The stainless-steel PST II was the second variation of the original Leatherman PST, which stands for Personal Survival Tool. This newer model had a diamond-coated file with a fishhook groove, a pair of scissors, and a half-serrated, half–straight edged knife blade. It is an art deco–like design of elegance and utility, a mix of Bond and MacGyver.

The Leatherman PST II fixed just about every problem. I told Koreans that it was my version of a *pujok*, a paper talisman that brings luck and wards off misfortune. My first apartment had a speaker mounted in the ceiling. Every morning, the caretaker of the apartment complex would cackle on at full volume. He gave a weather report and reminded people about garbage placement and encouraged everyone to work hard and be diligent. Sometimes he would talk at length about other things, but since my Korean wasn't very good at the time, I did not know what he was saying. His voice through the speaker was so loud that I could hear it in other apartments. It could not be turned off. I fixed this problem with the wire cutters on the Leatherman.

South Korea, like the rest of the world, is filled with many cheap consumer goods that easily break down. The tools of the Leatherman helped me handle those annoyances. I used the needle-nose pliers to remove countless fishhooks, and the knife blade prepped many shore lunches. I kept it sharp with a small Washita stone. South Korea is a country where the everyday carry of knives is looked upon as weird unless a blade is required for an occupational or recreational reason. Large knives are looked upon as tools of criminals unless they are in the hands of a butcher or a sashimi chef. The Leatherman was a discrete portable toolbox that saved my ass many times.

8. HAVE A PANIC HOLE FOR HOUSE FITS

Being an expatriate means temporary housing, usually an apartment. In Asia, it's probably an apartment in a high-rise surrounded by more high-rise apartment complexes. Apartment living in densely populated urban areas is suitable for some, but it never was for me. I always felt trapped and hemmed in. I self-diagnosed this agitation and merely wrote it off as cabin fever.

In Toni Morrison's novel *Beloved*, she mentions something similar, and when I first read it, I knew exactly what she was getting at: "He believed he was having house-fits, the glassy anger men sometimes feel when a woman's house begins to bind them, when they want to yell and break something or at least run off."

The solution is to get outside. But immediately outside that apartment is pure chaos: insane traffic, in-your-face advertising, masses of street people, and the grime and grunginess of urban living. For some, it's easier to stay in. The Japanese have the *hikikomori* (translated as "pulling inward" and "being confined"), and the South Koreans have the same subculture of social recluses that dwell in tiny apartments and don't go out. They are shut-ins, urban hermits.

I noticed that some expatriates I knew in South Korea became partial *hikikomori*. Other than going to work and shopping for basic needs, they holed up in their apartments until their schedule forced them back out. Home delivery of meals is fast and cheap in Korea, and many lived on that. One American guy I knew in Gwangyang overdosed on diazepam and whiskey and wasn't found until days later. It's like they were prisoners in cells of their own making. It's like Captain Willard in a Saigon hotel room but for months and years. Alcohol plays a part in this behavior, but the biggest influence on the creation of *hikikomori* is high-speed internet. Pull the plug on the online anchorites, and this social problem will quickly go away.

Author and poet Jim Harrison defines a "panic hole" in an essay of the same name: "Panic hole is defined as a place where you go physically or mentally or both when the life is being squeezed out of you or when you think it is, which is the same thing." In the essay, he was "feeling put upon, a close second to self-pity as a destructive state," and his panic hole was a cross-country road trip in a Toyota Land Cruiser.

For some, a panic hole is going to the gym for a workout or a temple for meditation. It could be an empty room in a building where you know you will not be bothered. For the *hikikomori*, the panic hole becomes a permanent safe space, which is an unnatural state of being. In South Korea, the establishment of the greenbelt made it possible for anyone to wander to the outskirts of a city, find a trail, and wander off into a mountain forest without any concern about trespassing. An outlaw always husbands a hideout, and my panic hole was to walk in the woods. The trail might lead up a steep mountain or into rolling hills. It might be pretty or it might be plain, but you are outside and alone and can peacefully deal with your troubled thoughts. *Solvitur ambulando* always worked for me.

9. EMBRACE THE LOCAL CUISINE BUT COOK WITH FIERCE INDEPENDENCE

Eat the local foods with curiosity and gusto. Eat as an omnivore and try everything. Go out and hit new places with friends and try different foods. This is enriching your life. There will be a time when you are craving a food that cannot be found where you are living. At a "Western" restaurant on Koje Island, I was once served spaghetti with ketchup used as marinara sauce and pizza (with ketchup) dotted with canned sweet corn.

Also, there will be a time when the cuisine of the host country will get repetitive. In Korean cooking, there isn't much of a tradition of baked goods, smoked meats, or charcuterie, any kind of dairy, or interesting desserts. This is easy to bitch about, but one of the pleasures I found was figuring out how to cook what you want with what you have. I often surprised my wife with cheesecake with a pie crust made of cracker crumbs from a favorite children's snack. I smoked fish in a metal garbage can. Breakfast carbonara was a staple in my home. Farmer's cheese congealed in my fridge. Egg-salad sandwiches hit the spot during fishing trips. My favorite was pickling Coturnix quail eggs. They were just like the pickled eggs you would find in a jar behind the bar in a backwoods tavern—just tinier versions.

10. REMEMBER WHERE YOU CAME FROM AND WHERE YOU ARE NOW

Some expatriates "go native" and embrace the local culture so much that it seems that they forgot their former identities. These types tend to shun fellow expatriates and live in their own world. Perhaps that is a good thing, and to each their own.

I spent so much time on the road and on the river that some weeks the only people I would speak to would be my wife, clients if I was guiding trips, and the *ajummas* at the corner stores when I would load up on supplies. One thing that helped me stay grounded was making it over to IP (the International Pub) in Changwon, a landmark watering hole owned and operated by Soon-young—a sweet little Korean lady who wore thick glasses and always greeted you with a smile. I would drink and talk with friends and acquaintances, freaks, soldiers, engineers, and sometimes world-class athletes. (The U.S. Olympic shooting team competed once in Changwon, and I spent an evening talking shotguns with them.) If I was passing through Seoul, I would sit in on a lecture by the Royal Asiatic Society, or I would swing into the Seoul Foreign Correspondents Club.

While living in a foreign country you can get so caught up in that world you forget the world where you came from. There were times when I did not know the names of current U.S. politicians, recent Hollywood films, or how the Green Bay Packers were doing. You must call home and stay in touch with old friends and family. Write both letters and emails. Remember birthdays. Celebrate holidays in both countries. Host a Christmas party. Embrace the country and culture you are living in but surround yourself with reminders of who you are and where you are from. I imported a Weber grill, and this became the centerpiece of my courtyard. On it I grilled Korean *samgyupsal* that became slow-smoked, thick-cut bacon for BPTs (bacon, perilla, tomato). This was cooked and eaten while listening to the Allman Brothers, and later we would retire to watch a Song Kang-ho movie. It's about mixing the best of both worlds.

IN THE OUTDOOR LIVES SERIES

*Pacific Lady: The First Woman to Sail
Solo across the World's Largest Ocean*
by Sharon Sites Adams
with Karen J. Coates

*Kayaking Alone: Nine Hundred Miles from
Idaho's Mountains to the Pacific Ocean*
by Mike Barenti

*The Dawn Patrol Diaries: Fly-Fishing
Journeys under the Korean DMZ*
by James Card

*Bicycling beyond the Divide:
Two Journeys into the West*
by Daryl Farmer

*Beneath Blossom Rain: Discovering
Bhutan on the Toughest Trek in the World*
by Kevin Grange

*Wilderness of Hope: Fly Fishing and
Public Lands in the American West*
by Quinn Grover

*The Hard Way Home: Alaska Stories of
Adventure, Friendship, and the Hunt*
by Steve Kahn

*Almost Somewhere: Twenty-Eight
Days on the John Muir Trail*
by Suzanne Roberts

*Stories from Afield: Adventures
with Wild Things in Wild Places*
by Bruce L. Smith

*Beautifully Grotesque Fish
of the American West*
by Mark Spitzer

To order or obtain more information on these or other University
of Nebraska Press titles, visit nebraskapress.unl.edu.

Printed in the USA
CPSIA information can be obtained
at www.ICGtesting.com
CBHW020340170824
13306CB00006B/309

9 781496 234490